Advances in Electrochemical Energy Materials

Advances in Electrochemical Energy Materials

Special Issue Editors

Zhaoyang Fan
Shiqi Li

MDPI • Basel • Beijing • Wuhan • Barcelona • Belgrade

Special Issue Editors
Zhaoyang Fan
Texas Tech University
USA

Shiqi Li
Hangzhou Dianzi University
China

Editorial Office
MDPI
St. Alban-Anlage 66
4052 Basel, Switzerland

This is a reprint of articles from the Special Issue published online in the open access journal *Materials* (ISSN 1996-1944) from 2018 to 2020 (available at: https://www.mdpi.com/journal/materials/special_issues/electrochemical_energy_materials).

For citation purposes, cite each article independently as indicated on the article page online and as indicated below:

LastName, A.A.; LastName, B.B.; LastName, C.C. Article Title. *Journal Name* **Year**, *Article Number*, Page Range.

ISBN 978-3-03928-642-3 (Hbk)
ISBN 978-3-03928-643-0 (PDF)

Cover image courtesy of Wenyue Li.

© 2020 by the authors. Articles in this book are Open Access and distributed under the Creative Commons Attribution (CC BY) license, which allows users to download, copy and build upon published articles, as long as the author and publisher are properly credited, which ensures maximum dissemination and a wider impact of our publications.

The book as a whole is distributed by MDPI under the terms and conditions of the Creative Commons license CC BY-NC-ND.

Contents

About the Special Issue Editors vii

Shiqi Li and Zhaoyang Fan
Special Issue: Advances in Electrochemical Energy Materials
Reprinted from: *Materials* **2020**, *13*, 844, doi:10.3390/ma13040844 1

Wen Zhang, Junfan Zhang, Yan Zhao, Taizhe Tan and Tai Yang
High Electrochemical Performance of Nanotube Structured ZnS as Anode Material for Lithium–Ion Batteries
Reprinted from: *Materials* **2018**, *11*, 1537, doi:10.3390/ma11091537 5

Chengkang Chang, Jian Dong, Li Guan and Dongyun Zhang
Enhanced Electrochemical Performance of $Li_{1.27}Cr_{0.2}Mn_{0.53}O_2$ Layered Cathode Materials via a Nanomilling-Assisted Solid-state Process
Reprinted from: *Materials* **2019**, *12*, 468, doi:10.3390/ma12030468 14

Jun Liu, Qiming Liu, Huali Zhu, Feng Lin, Yan Ji, Bingjing Li, Junfei Duan, Lingjun Li and Zhaoyong Chen
Effect of Different Composition on Voltage Attenuation of Li-Rich Cathode Material for Lithium-Ion Batteries
Reprinted from: *Materials* **2020**, *13*, 40, doi:10.3390/ma13010040 27

Qiming Liu, Huali Zhu, Jun Liu, Xiongwei Liao, Zhuolin Tang, Cankai Zhou, Mengming Yuan, Junfei Duan, Lingjun Li and Zhaoyong Chen
High-Performance Lithium-Rich Layered Oxide Material: Effects of Preparation Methods on Microstructure and Electrochemical Properties
Reprinted from: *Materials* **2020**, *13*, 334, doi:10.3390/ma13020334 39

Zhiyong Yu, Jishen Hao, Wenji Li and Hanxing Liu
Enhanced Electrochemical Performances of Cobalt-Doped Li_2MoO_3 Cathode Materials
Reprinted from: *Materials* **2019**, *12*, 843, doi:10.3390/ma12060843 51

Rongyue Liu, Jianjun Chen, Zhiwen Li, Qing Ding, Xiaoshuai An, Yi Pan, Zhu Zheng, Minwei Yang and Dongju Fu
Preparation of $LiFePO_4$/C Cathode Materials via a Green Synthesis Route for Lithium-Ion Battery Applications
Reprinted from: *Materials* **2018**, *11*, 2251, doi:10.3390/ma11112251 62

Abhishek Sarkar, Pranav Shrotriya and Abhijit Chandra
Simulation-driven Selection of Electrode Materials Based on Mechanical Performance for Lithium-Ion Battery
Reprinted from: *Materials* **2019**, *12*, 831, doi:10.3390/ma12050831 75

Yan Ji, Cankai Zhou, Feng Lin, Bingjing Li, Feifan Yang, Huali Zhu, Junfei Duan and Zhaoyong Chen
Submicron-Sized Nb-Doped Lithium Garnet for High Ionic Conductivity Solid Electrolyte and Performance of Quasi-Solid-State Lithium Battery
Reprinted from: *Materials* **2020**, *13*, 560, doi:10.3390/ma13030560 89

Dongya Sun, Liwen He, Yongle Lai, Jiqiong Lian, Jingjing Sun, An Xie and Bizhou Lin
Structure and Electrochemical Properties of Mn_3O_4 Nanocrystal-Coated Porous Carbon Microfiber Derived from Cotton
Reprinted from: *Materials* **2018**, *11*, 1987, doi:10.3390/ma11101987 100

Roger Amade, Arevik Muyshegyan-Avetisyan, Joan Martí González, Angel Pérez del Pino, Eniko György, Esther Pascual, José Luís Andújar and Enric Bertran Serra
Super-Capacitive Performance of Manganese Dioxide/Graphene Nano-Walls Electrodes Deposited on Stainless Steel Current Collectors
Reprinted from: *Materials* **2019**, *12*, 483, doi:10.3390/ma12030483 107

Rabia Ahmad, Naseem Iqbal and Tayyaba Noor
Development of ZIF-Derived Nanoporous Carbon and Cobalt Sulfide-Based Electrode Material for Supercapacitor
Reprinted from: *Materials* **2019**, *12*, 2940, doi:10.3390/ma12182940 120

Wenyue Li, Nazifah Islam, Guofeng Ren, Shiqi Li and Zhaoyang Fan
AC-Filtering Supercapacitors Based on Edge Oriented Vertical Graphene and Cross-Linked Carbon Nanofiber
Reprinted from: *Materials* **2019**, *12*, 604, doi:10.3390/ma12040604 131

About the Special Issue Editors

Zhaoyang Fan (Professor) obtained his B.E. and M.E. degrees from Tsinghua University of China and Ph.D. from Northwestern University of the U.S. He is a professor in the Department of Electrical and Computer Engineering, Texas Tech University. His research concerns wide bandgap semiconductor materials and devices, nanomaterials, and electrochemical energy storage.

Shiqi Li (Associate Professor) received his B.S. from Wuhan University in 2005 and Ph.D. from Peking University in 2010. After several years in the industry and in the academy, he was appointed as an associate professor at Hangzhou Dianzi University in 2017. His interest focuses on electrochemical energy storage, particularly Li-S batteries, Li metal anode, and supercapacitors.

Editorial

Special Issue: Advances in Electrochemical Energy Materials

Shiqi Li [1] and Zhaoyang Fan [2,*]

1. College of Electronic Information, Hangzhou Dianzi University, Hangzhou 310018, China; sqli@hdu.edu.cn
2. Department of Electrical and Computer Engineering and Nano Tech Center, Texas Tech University, Lubbock, TX 79409, USA
* Correspondence: zhaoyang.fan@ttu.edu

Received: 9 February 2020; Accepted: 11 February 2020; Published: 13 February 2020

Abstract: Electrochemical energy storage is becoming essential for portable electronics, electrified transportation, integration of intermittent renewable energy into grids, and many other energy or power applications. The electrode materials and their structures, in addition to the electrolytes, play key roles in supporting a multitude of coupled physicochemical processes that include electronic, ionic, and diffusive transport in electrode and electrolyte phases, electrochemical reactions and material phase changes, as well as mechanical and thermal stresses, thus determining the storage energy density and power density, conversion efficiency, performance lifetime, and system cost and safety. Different material chemistries and multiscale porous structures are being investigated for high performance and low cost. The aim of this Special Issue is to report the recent advances of materials used in electrochemical energy storage that encompasses supercapacitors and rechargeable batteries.

Keywords: lithium ion batteries; supercapacitors; electrode materials; nanostructure; electrochemical energy storage

---

Electrochemical energy materials are used for electrochemical energy storage or conversion. Broadly speaking, these include materials used in batteries and supercapacitors, as well as electrocatalysts to produce new fuels. In this Special Issue, we focus on those in lithium-ion batteries (LIBs) and supercapacitors, particularly the electrode active materials and their structure that must be capable of supporting multitude of coupled physicochemical processes as well as mechanical and thermal stresses. They directly determine the overall performance of the energy storage, including ultimate energy and power densities, lifetime, and system cost and safety.

The commercialized LIB now uses graphitic carbon as its anode, which has a theoretical capacity of 372 mAh g^{-1} based on Li^+ intercalation between the graphite layers. Many other materials can form alloys with lithium and thus provide much higher capacity. These materials, however, generally suffer from a large volume change during the alloying-dealloying process, leading to quick fading of the anode capacity. Nanostructure engineering is a practical approach to release stress, thus minimizing electrode material pulverization. In contrast to the anode, the overall performance of a LIB nowadays is largely constrained by its cathode, which has only about half the specific capacity of the graphite anode. The cathode is also the most expensive and the heaviest component in an LIB. Therefore, increasing the cathode specific capacity is crucial for better and cheaper LIBs. In this regard, Li-rich manganese-based layered oxides with the chemical formula $xLi_2MnO_3 \cdot (1-x)LiMO_2$, where M represents transition metal elements, with its capacity up to 300 mAh g^{-1} has drawn considerable attention. Other cathode materials also still have enough room for further improvement of their performance. In addition to electrodes, developing solid electrolytes in substitution of the liquid electrolyte and the separator for producing solid-state LIBs is another active area. The solid-state LIBs will address the safety issue related to the organic solvent-based electrolyte and also have potential to increase the energy density.

LIBs can store a large amount of energy, but the slow kinetics in the electrochemical process restrain the rate of energy storing and releasing, or the charging current rate and output power density. There are plenty of applications that require high-power and high-rate energy storage with much long cycle lifetime where LIBs cannot meet the demand. Electrochemical supercapacitors can fit into these needs very well.

Conventional supercapacitors are those storing charges electrostatically in the electrical double layer formed on an inertial carbon surface, or electrical double layer capacitors (EDLCs). They offer a high-power density and a long cycle lifetime, but with an energy density less than 10 Wh kg^{-1}, more than 20 folds smaller than that of LIBs. Therefore, pseudocapacitors are being actively investigated to store charges in a surface-related reversible faradic redox reaction, thus offering much larger capacitance than EDLCs to bridge the energy density gap from batteries. On the other end of the spectrum, EDLC has a frequency response limited to 1 Hz or so, mainly caused by its mesoporous carbon electrode structure. Developing ultrafast EDLCs that can effectively work at hundreds and even kHz domain will broaden the function of EDLCs into the area of filtering capacitors and therefore, is also attracting much attention.

The Special Issue "Advances in Electrochemical Energy Materials" was proposed to present recent developments in this active field. The twelve articles included touch different aspects of materials for electrochemical energy storage, which are introduced in the following.

The main theme of material research for LIBs is centred on the high capacity anode and cathode. The article by Zhang et al. [1] reported on nanotube structured ZnS as the anode of LIBs. ZnS is considered as a promising alternative to graphitic carbon due to its much higher theoretical capacity of 962.3 mAh g^{-1}, but its large volume change during the charging and discharging process hinders its practical application. Nanotube structured ZnS anode was therefore prepared using ZnO nanotubes as a sacrificial template, expecting that the radially and longitudinally expansion of nanotubes could mitigate the stress and thus improving the electrode stability. A high initial capacity and reasonable cycling stability were demonstrated for this ZnS anode structure.

Developing high-capacity and low-cost cathode materials for LIBs has attracted considerable attention. This is particularly true to Li-rich layered oxides. However, these materials suffer from severe voltage and capacity fading caused by continuous phase transition from layered phase to spinel or others during the repeated Li^+ extraction/insertion process. Composition control to maintain the structural stability is hence crucial for cathodes with prolonged structural integrity and enhanced electrochemical performance. Chang et al. [2] reported a study of layered $Li_{1.27}Cr_{0.2}Mn_{0.53}O_2$ powders with mesoporous structure synthesized by a nanomilling-assisted solid-phase method. The fabricated cathode delivered a capacity close to its theoretical value with good capacity retention after 100 charging-discharging cycles. No transformation of the layered crystal structure was confirmed. Two papers from Chen's group [3,4] presented their studies on the Li-rich manganese-based layered oxides. It was found that a high nickel content in the layered phase could stabilize the structure and alleviate the voltage and capacity attenuation [3]. This was explained that some Ni^{2+} ions occupy the Li^+ ion sites and this cation doping improves the structural stability by supporting the Li slabs and reducing tension of neighboring oxygen layers during the delithiation process. The preferential reduction of $Ni^{4+/2+}$ also maintains the average oxidation state of Mn above 3^+, effectively improving structural durability. Composition uniformity is another crucial parameter which might be related to the synthesis method [4]. The sol–gel and the oxalate co-precipitation synthesis methods were subsequently compared based on the microstructure, element distribution, and electrochemical performance of the prepared manganese oxides with a high nickel content. The uniform element distribution in samples synthesized by the oxalate co-precipitation method further contributed to the stability of the layered structure.

Other than manganese-based, Li-rich molybdenum-based layered oxides was also attractive. Yu et al. [5] investigated Co doping in Li_2MoO_3 to improve its structure stability and electronic conductivity. Their results showed that an appropriate amount of Co ions can be introduced into the Li$_2$MoO$_3$ lattices and electrochemical tests revealed that Co-doping can significantly improve the electrochemical performances of the Li$_2$MoO$_3$ materials.

In addition to these new cathode materials, a further study of the conventional olivine-type LiFePO$_4$, was also carried out, aiming to reduce the manufacturing cost and minimize pollutants generation. Liu et al. [6] developed a green route to produce the LiFePO$_4$/C composite, which showed a uniform carbon coating on LiFePO$_4$ nanoparticles, with effectively improved conductivity and enhanced Li$^+$ ion diffusion. Consequently, LIBs using the synthesized composite as cathode materials exhibited superior performance, especially at high rates.

Besides experimental test of electrode stability, simulation-driven selection of electrode materials based on mechanical performance during lithiation/delithiation process was also studied. Sarkar et al. [7] developed a model to determine particle deformation and stress fields by combining the stress equilibrium equations with the Li$^+$ electrochemical diffusion. It was applied to derive five merit indices to reflect the mechanical stability of electrode materials. The authors further suggested ways for the selection and optimal design of electrode materials to improve their mechanical performance.

Solid-state LIBs are being pursued as the next-generation energy storage technology to provide high safety and high energy density. For this technology, the solid electrolyte with high ionic conductivity and electrochemical stability is the most crucial component. Ji et al. [8] studied the synthesis of Nb-doped lithium garnet Li$_7$La$_3$Zr$_2$O$_{12}$ (LLZNO) as a high ionic conductive solid electrolyte. Submicron size LLZNO powder was prepared using a solid-state reaction and an attrition milling process, followed by sintering at a relatively low temperature for a short time. The properties of the synthesized LLZNO and its performance in a solid-state LIB were reported.

This Special Issue also includes several papers presenting the research on electrode materials and structures for supercapacitors, particularly for pseudocapacitors. The electrode materials for pseudocapacitors commonly include transition metal oxides, nitrides, and sulfides, among others. Since the pseudocapacitive effect is commonly surface or sub-surface related, a large surface area of these compounds is crucial for achieving a high specific capacitance. Their generally low conductivity is another issue to be addressed for high-rate and high-power performance. These compounds, therefore, are commonly synthesized into a nanoparticle form anchored on a carbon-based conductive framework. In the work by Sun et al. [9], a biomorphic porous composite was prepared with Mn$_3$O$_4$ nanocrystals anchored on porous carbon microfiber, with the latter derived from cotton wool. The unique structure resulted in the good cycling stability of the fabricated supercapacitors. Amade et al. [10] reported using graphene nanowalls, which were grown in a plasma-enhanced chemical vapor deposition process, as the framework for manganese dioxide deposition by electrodeposition. More interesting work by Ahmad et al. [11] investigated nanoporous carbon, derived from zeolitic imidazolate framework (ZIF-67) as the support of cobalt sulfide, which was formed through anion exchange sulfidation process from cobalt oxide. A large capacitance of 677 F g^{-1} was obtained.

There is strong interest in developing high-frequency supercapacitors or electrochemical capacitors (HF-ECs) [12] for line-frequency alternating current (AC) filtering in the substitution of bulky aluminum electrolytic capacitors, with broad applications in the power and electronic fields. Edge-oriented vertical graphene networks on 3D scaffolds have a unique structure that offers straightforward pore configuration, reasonable surface area, and high electronic conductivity, thus allowing the fabrication of HF-ECs. Comparatively, highly conductive freestanding cross-linked carbon nanofibers, derived from bacterial cellulose in a rapid plasma pyrolysis process can also provide a large surface area but are free of rate-limiting micropores, and are another good candidate for HF-ECs. Li et al. [12] summarized the recent advances in this field with emphasis on their contributions in the study of these materials and their electrochemical properties including preliminary demonstrations of HF-ECs for AC line filtering and pulse power storage applications.

Funding: This research received no external funding.

Conflicts of Interest: The authors declare no conflict of interest.

References

1. Zhang, W.; Zhang, J.; Zhao, Y.; Tan, T.; Yang, T. High Electrochemical Performance of Nanotube Structured ZnS as Anode Material for Lithium–Ion Batteries. *Materials* **2018**, *11*, 1537. [CrossRef] [PubMed]
2. Chang, C.; Dong, J.; Guan, L.; Zhang, D. Enhanced Electrochemical Performance of $Li_{1.27}Cr_{0.2}Mn_{0.53}O_2$ Layered Cathode Materials via a Nanomilling-Assisted Solid-state Process. *Materials* **2019**, *12*, 468. [CrossRef] [PubMed]
3. Liu, J.; Liu, Q.; Zhu, H.; Lin, F.; Ji, Y.; Li, B.; Duan, J.; Li, L.; Chen, Z. Effect of Different Composition on Voltage Attenuation of Li-Rich Cathode Material for Lithium-Ion Batteries. *Materials* **2020**, *13*, 40. [CrossRef] [PubMed]
4. Liu, Q.; Zhu, H.; Liu, J.; Liao, X.; Tang, Z.; Zhou, C.; Yuan, M.; Duan, J.; Li, L.; Chen, Z. High-Performance Lithium-Rich Layered Oxide Material: Effects of Preparation Methods on Microstructure and Electrochemical Properties. *Materials* **2020**, *13*, 334. [CrossRef] [PubMed]
5. Yu, Z.; Hao, J.; Li, W.; Liu, H. Enhanced Electrochemical Performances of Cobalt-Doped Li_2MoO_3 Cathode Materials. *Materials* **2019**, *12*, 843. [CrossRef] [PubMed]
6. Liu, R.; Chen, J.; Li, Z.; Ding, Q.; An, X.; Pan, Y.; Zheng, Z.; Yang, M.; Fu, D. Preparation of $LiFePO_4$/C Cathode Materials via a Green Synthesis Route for Lithium-Ion Battery Applications. *Materials* **2018**, *11*, 2251. [CrossRef] [PubMed]
7. Sarkar, A.; Shrotriya, P.; Chandra, A. Simulation-driven Selection of Electrode Materials Based on Mechanical Performance for Lithium-Ion Battery. *Materials* **2019**, *12*, 831. [CrossRef] [PubMed]
8. Ji, Y.; Zhou, C.; Lin, F.; Li, B.; Yang, F.; Zhu, H.; Duan, J.; Chen, Z. Submicron-Sized Nb-Doped Lithium Garnet for High Ionic Conductivity Solid Electrolyte and Performance of Quasi-Solid-State Lithium Battery. *Materials* **2020**, *13*, 560. [CrossRef] [PubMed]
9. Sun, D.; He, L.; Lai, Y.; Lian, J.; Sun, J.; Xie, A.; Lin, B. Structure and Electrochemical Properties of Mn_3O_4 Nanocrystal-Coated Porous Carbon Microfiber Derived from Cotton. *Materials* **2018**, *11*, 1987. [CrossRef] [PubMed]
10. Amade, R.; Muyshegyan-Avetisyan, A.; Martí González, J.; Pérez del Pino, A.; György, E.; Pascual, E.; Andújar, J.L.; Bertran Serra, E. Super-capacitive performance of manganese dioxide/graphene nano-walls electrodes deposited on stainless steel current collectors. *Materials* **2019**, *12*, 483. [CrossRef] [PubMed]
11. Ahmad, R.; Iqbal, N.; Noor, T. Development of ZIF-Derived Nanoporous Carbon and Cobalt Sulfide-Based Electrode Material for Supercapacitor. *Materials* **2019**, *12*, 2940. [CrossRef] [PubMed]
12. Li, W.; Islam, N.; Ren, G.; Li, S.; Fan, Z. AC-Filtering Supercapacitors Based on Edge Oriented Vertical Graphene and Cross-Linked Carbon Nanofiber. *Materials* **2019**, *12*, 604. [CrossRef] [PubMed]

© 2020 by the authors. Licensee MDPI, Basel, Switzerland. This article is an open access article distributed under the terms and conditions of the Creative Commons Attribution (CC BY) license (http://creativecommons.org/licenses/by/4.0/).

Article

High Electrochemical Performance of Nanotube Structured ZnS as Anode Material for Lithium–Ion Batteries

Wen Zhang [1], Junfan Zhang [1], Yan Zhao [1,*], Taizhe Tan [2] and Tai Yang [1,*]

[1] School of Materials Science and Engineering, Research Institute for Energy Equipment Materials, Hebei University of Technology, Tianjin 300130, China; zhangwen@hebut.edu.cn (W.Z.); 18722593259@163.com (J.Z.)
[2] Synergy Innovation Institute of GDUT, Heyuan 517000, Guangdong, China; tztansii18@163.com
* Correspondence: yanzhao1984@hebut.edu.cn (Y.Z.); yangtai999@163.com (T.Y.); Tel.: +86-22-6020-1433 (T.Y.)

Received: 3 July 2018; Accepted: 23 August 2018; Published: 26 August 2018

Abstract: By using ZnO nanorods as an ideal sacrificial template, one-dimensional (1-D) ZnS nanotubes with a mean diameter of 10 nm were successfully synthesized by hydrothermal method. The phase composition and microstructure of the ZnS nanotubes were characterized by using XRD (X-ray diffraction), SEM (scanning electron micrograph), and TEM (transmission electronic microscopy) analysis. X-ray photoelectron spectroscopy (XPS) and nitrogen sorption isotherms measurements were also used to study the information on the surface chemical compositions and specific surface area of the sample. The prepared ZnS nanotubes were used as anode materials in lithium-ion batteries. Results show that the ZnS nanotubes deliver an impressive prime discharge capacity as high as 950 mAh/g. The ZnS nanotubes also exhibit an enhanced cyclic performance. Even after 100 charge/discharge cycles, the discharge capacity could still remain at 450 mAh/g. Moreover, cyclic voltammetry (CV) and electrochemical impedance spectroscopy (EIS) measurements were also carried out to evaluate the ZnS electrodes.

Keywords: lithium-ion batteries; zinc sulfide; nanotubes; anode material; electrochemical performance

1. Introduction

In recent decades, lithium-ion batteries play an increasingly dominating role in portable electronic devices due to the fact that they have the advantages of long service life, high energy density, high reversible capacity, and environmental friendliness [1]. Graphitic materials as a conventional anode material in lithium-ion batteries are extensively used for their good electrochemical properties and their structural stability during charge–discharge cycling [2]. However, traditional graphitic carbon materials severely hinder the development of lithium-ion batteries due to their low theoretical capacity (372 mAh/g) [3,4]. In order to meet energy storage needs, it is necessary to exploit new types of anode materials to replace carbon materials. Metal sulfides—such as CuS, MoS_2, NiS, and ZnS—also have been used as anodic electrode materials in lithium-ion batteries [5–8]. For example, CuS/graphene composite have a good charge–discharge cycling performance; however, its initial discharge capacity was only 627 mAh/g [9]. NiS-carbon nanofiber films have worse electrochemical properties, and its discharge capacity decayed below 100 mAh/g after 40 cycles [10]. MoS_2 nanowall/graphene has a stable discharge capacity of about 700 mAh/g [11], which is still unimpressive. ZnS, by contrast, is viewed as a very promising alternative to carbon anode material due to its high theoretical capacity (962.3 mAh/g) [12]. Unfortunately, some drawbacks hinder its commercialization process. The main problem is significant volume changes during its charging and discharging processes, which leads to a great capacity fade upon cycling [13]. Moreover, normal ZnS particles have poor electrical conductivity, as a result, anode electrodes made using unmodified ZnS suffer from a poor cyclic and rate performance [14].

There are two possible ways to solve above mentioned problems. For one thing, we should focus on the synthesis of nano-sized particles, which can effectively adapt to volume changes during the charge–discharge progress [15–17]. For another, it is an effective method to combine nano-structured ZnS particles with conductive carbon coating to increase conductivity of anodic materials in lithium-ion batteries [18]. He et al. [14] prepared ZnS/C composites by a combined precipitation with carbon coating method and applied them as anode material for lithium-ion batteries. Du et al. [12] also synthesized nanocrystalline ZnS/C with core/shell structure by using a simple solvothermal process and an annealing process. These studies have made some progress in development of anodic materials for lithium-ion batteries. Nevertheless, the above-mentioned preparation methods of nanocrystalline ZnS/C composites are complicated and costly. 1-D ZnS nanotubes also have been proved to be a promising candidate material [19,20]. It is well-known that active materials in anodes with large surface areas can increase the contact area between electrolyte and electrode materials, thereby enhancing energy storage density [21]. Moreover, nano-materials can also shorten the transport path of conductive ions, and the electrodes will not be destroyed even though a large volume change of ZnS occurs in the charge–discharge process [22,23]. It was reported that nanotubes can expand radially as well as longitudinally to mitigate the stress, which would make them more suitable for high rate applications [24].

In order to further investigate the electrochemical performance of nano-structured sulfides, ZnS nanotubes were prepared by hydrothermal method by using ZnO nanorod arrays as sacrificial template. The prepared ZnS nanotubes exhibit a well rate discharge performance. The discharge capacity of ZnS nanotubes is as high as 950 mAh/g in the first cycle, and it still remains at 450 mAh/g after 100 charge/discharge cycles.

2. Materials and Methods

2.1. Synthesis of ZnS Samples

Firstly, 25 mmol of $Zn(NO_3)_2 \cdot 6H_2O$ and 50 mL of polyvinyl pyrrolidone (PVP) aqueous solution (0.1 wt %) were mixed with a certain amount of deionized water to get $Zn(NO_3)_2$ solution of 0.05 mol/L. At the same time, hexamethylenetetramine ($C_6H_{12}N_4$) solution (50 mL, 0.1 wt %) was also prepared. Above two solutions were mixed, heated, and stirred in a beaker-flask at 90 °C for 16 h. Then the white products were collected and washed by deionized water and ethanol three times, and the precipitate was dried in a vacuum oven at 70 °C for 12 h. Finally, pure ZnO nanorods were obtained.

Subsequently, the ZnS nanotubes were synthesized by hydrothermal method by using the ZnO nanorods as template. The prepared pure ZnO nanorods were dispersed in 20 mL of ethylene glycol ($C_2H_6O_2$) solution, stirring and sonicating for 20 min. After that, thioacetamide (CH_3CSNH_2) was dripped into the above-mentioned ZnO suspension. The mixture solution was transferred to a Teflon-lined stainless-steel autoclave and placed into an oven maintained at 145 °C for 10 h. After this reaction, the ZnO/ZnS nano composites were collected and washed three times using deionized water and pure ethanol. Then, 2 g of ZnO/ZnS nano composites were added into 50 mL of 10 M NaOH aqueous solution and stirred for 2 h at room temperature to remove ZnO cores. The products were collected and dried at 80 °C for 10 h, and then white ZnS nanotubes were obtained.

2.2. Sample Characterizations

XRD method was used to analyze the phase composition and crystal structure of the sample. The tests were performed at a scanning rate of 2°/min in the 2θ range from 20° to 90° by using an X-ray diffractometer (SmartLab Rigaku Corporation, Tokyo, Japan). Identification of the species was computer aided. The microstructure and corresponding selected area electron diffraction (SAED) patterns for the ZnS nanotubes were also performed by using SEM (Hitachi S-4800) and TEM (JEOL-2010). Nitrogen sorption isotherms and Brunauer–Emmett–Teller (BET) surface area were measured at 423 K with a

V-Sorb 2800P analyzer (GAPP, Beijing, China). XPS (Thermo Fisher Scientific, Waltham, MA, USA) measurements were conducted to evaluate the chemical states of elements in the sample.

2.3. Electrochemical Measurements

The electrochemical behaviors of the ZnS nanotubes were characterized by using CR 2025 coin cell. In order to prepare working electrodes (anodic electrodes), a slurry was mixed by using 70 wt % of ZnS nanotube powder, 15 wt % of carbon black and 15 wt % of polyvinylidene fluoride. The mixture was grinded for 40 min and dissolved in N-methyl-2-pyrrolidone (NMP), and the obtained slurry with a thickness of 0.1 mm was blade cast onto Cu foil. Then the prepared electrode material was dried at 70 °C for 12 h. After that, the dried electrodes were punch into coins in an argon-filled (99.999%) glove box. The ZnS loading amount of each electrode sheet was approximately 2.5 mg/cm^2. Pure lithium metal foils were used as reference anode, and microporous polypropylene as a separator. The electrolyte was a solution of 1 mol/L LiPF$_6$ in ethylene carbonate ($C_3H_4O_3$) and dimethyl carbonate ($C_3H_6O_3$) with a volume ratio of 1:1. The charging and discharging measurements and cycle life tests of the prepared coin cells were carried out by using a multichannel battery testing system (Neware BTS4000). Considering the theoretical capacity (962.3 mAh/g) of ZnS [12], the charge–discharge current density of 962.3 mA/g was defied as 1 C. After the 100th charge–discharge cycle, the cells were dismantled to collect the anode materials. Then the anode materials were soaked in N-methyl-2-pyrrolidone (NMP) for 4 h to remove the binder and conductive agent. The phase structure and micro morphology of the collected ZnS nanotubes were also carried out by XRD and SEM. The charge–discharge voltage ranged from 0.05 V to 3.00 V. CV curves for the first three cycles were performed by an electrochemical workstation (Princeton, Versa STAT 4) at a scan rate of 0.1 mV/s in a voltage range of 0.01–3.00 V. The EIS measurements were also performed by the same electrochemical workstation with a frequency range of 10 kHz–10 mHz with a small sinusoidal perturbation of 10 mV.

3. Results and Discussion

3.1. Structural and Composition Characterization

SEM and TEM analysis were used to clarify the fine microstructures and morphologies of the ZnS sample. Figure 1a,b shows the SEM images of the ZnS nanotubes. It can be easily observed that the morphology of the ZnS sample is a kind of hollow micro tube, with the tube wall thickness of about 80 nm and the length was 1–2 µm. Detailed structural information of the ZnS sample was further investigated by TEM, results are shown in Figure 1c,d. Clearly, the nanotubes have a rough surface. The SAED pattern confirms the existence of ZnS. The three bright ED patterns correspond to the (111), (220), and (311) lattice plane of ZnS. Moreover, it can be observed from the high-resolution image shown in Figure 1d that the nanotubes are mainly composed of nanocrystals. This type of nanostructure contributes to the enhancement of electrochemical performance for the electrodes.

The XRD pattern of the synthesized ZnS nanotube sample and corresponding JCPDS data are shown in Figure 2a. Sharp diffraction peaks indicate good crystallinity of the sample. All of the diffraction peaks correspond well with the data of ZnS (JCPDS no. 65-0309). The three major diffraction peaks located at 2θ = 28.5°, 47.5°, and 56.3° correspond to (111), (220), and (311) crystal planes of ZnS. In order to know actual surface area of the ZnS nanotubes, N$_2$ adsorption and desorption isotherms are carried out, results are shown in Figure 2a. Type IV isotherm curve is observed with hysteresis loop at higher pressure, indicating a large number of meso-pores present in the sample [25]. The BET specific surface area of the sample was as high as 86.86 m^2/g. The XPS measurement was also conducted to obtain the information on the surface chemical compositions and the valence states of corresponding elements in the sample. From Figure 2c, it can be seen that the XPS spectra of S 2p was divided into two peaks centered at 163.1 and 162.0 eV, corresponding to S 2p$_{1/2}$ and S 2p$_{3/2}$ states [26]. Figure 2d depicted the XPS spectrum of the Zn 2p peaks centered at 1044.2 and 1021.3 eV, which associated with Cu 2p$_{1/2}$ and Cu 2p$_{3/2}$, respectively [27].

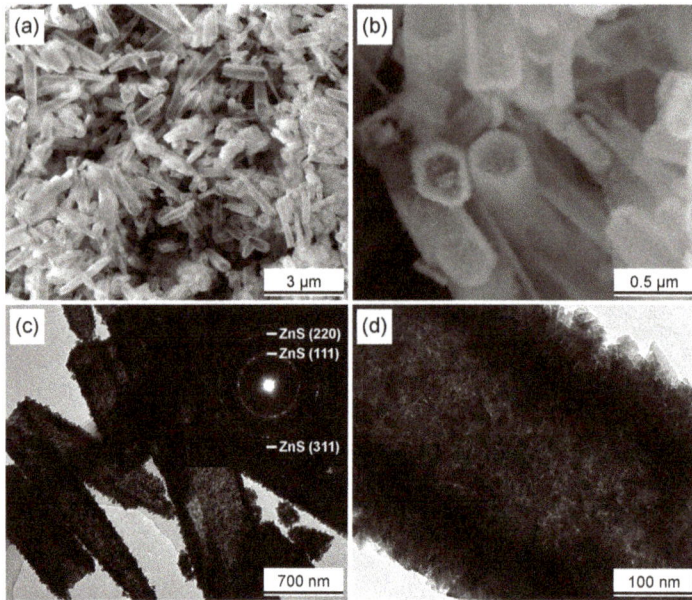

Figure 1. (**a**,**b**) SEM images of the prepared ZnS nanotubes; (**c**,**d**) TEM images together with corresponding SAED patterns of the ZnS nanotubes.

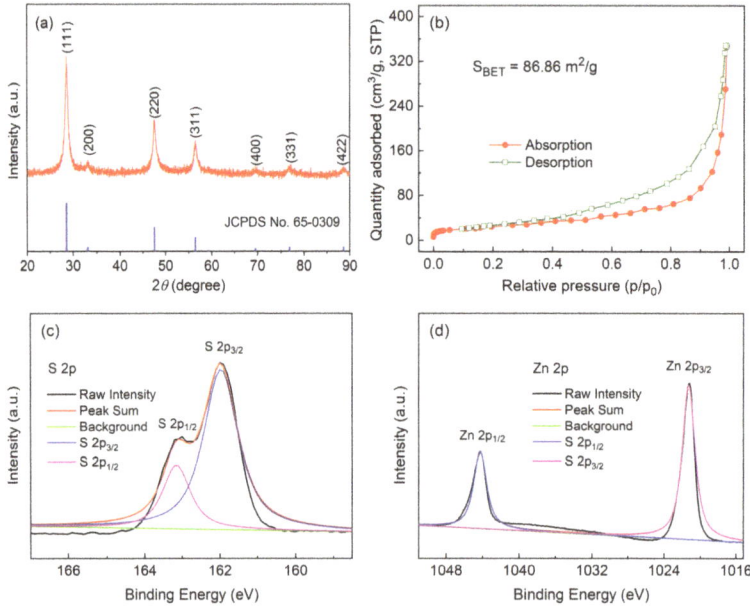

Figure 2. (**a**) XRD pattern of the ZnS nanotubes; (**b**) N_2 adsorption-desorption isotherm of the ZnS nanotubes; (**c**,**d**) XPS analysis for the ZnS nanotubes.

3.2. Electrochemical Performance

The discharge rate capabilities of the ZnS electrodes were tested at different current densities, as shown in Figure 3a. It can be observed that the discharge capacity of the ZnS electrode was steadily and has the high reversible capacities of 610, 500, 410, and 320 mAh/g at the discharge current density of 0.2, 0.5, 1, and 2 C. The specific capacity of the electrode is nearly recovered to its initial value in the case of the current density is goes back abruptly from 2 C to 0.2 C, further proving good reversibility and excellent cycling stability of the ZnS nanotube electrodes. The good rate discharge performance of the ZnS nanotubes is mainly due to the shortened lithium-ion diffusion distance and enhanced structural stability of the unique nanotubular structure [28]. Moreover, the hollow tubular structure can greatly increase the surface area of the material, which increases contact area between electrolyte and ZnS electrode material [29].

The charge–discharge profiles of the ZnS electrodes were evaluated by using galvanostatic method at the current density of 0.2 C, as shown in Figure 3b. During the initial discharge process, a typical slope can be clearly seen when the voltage was higher than 0.5 V. A major discharge voltage plateau can be observed at about 0.45 V, which represents to the lithiation reaction of ZnS nanotubes [30]. In the next few cycles, several typical charge/discharge stages can be observed, which could be due to the electrochemical activation of the system and tend to be stable [31]. The capacity fade during the first cycle can be result from the decomposition of electrolyte, which leads to the redistribution of the active materials [32]. In the following charge–discharge process, the ZnS electrode exhibits the same curves, indicating that the ZnS electrode has become stable. According to He et al. [14], the lithium insertion/extraction mechanism of ZnS active electrode material can be expressed: $(x + 2)Li^+ + ZnS + (x + 2)e^- \leftrightarrow Li_2S + Li_xZn$.

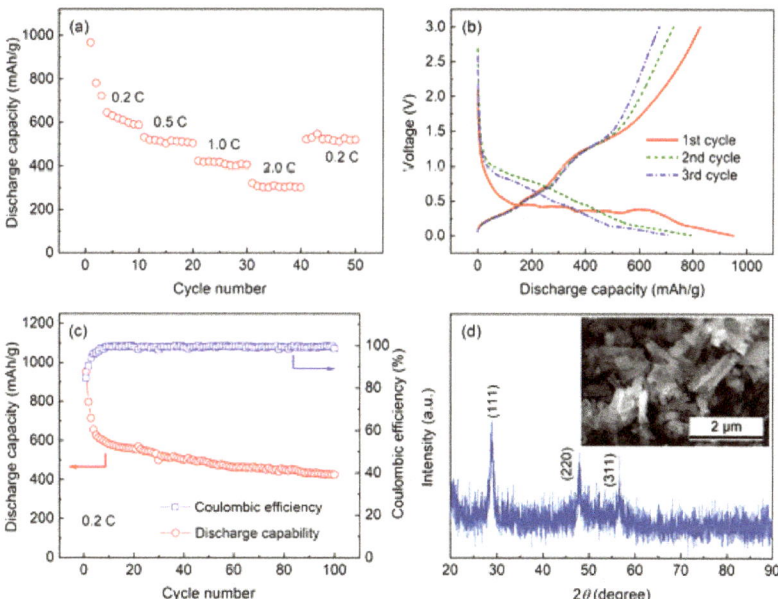

Figure 3. (a) Rate discharge capability of the ZnS electrodes together with the SEM images of ZnS nanotubes; (b) the first three galvanostatic charge/discharge profiles of the ZnS electrodes; (c) cycling performance and coulombic efficiency of the ZnS electrodes; (d) XRD pattern and SEM images of the ZnS nanotubes after 100th charge–discharge cycles.

Figure 3c presents the cycling performance of the ZnS electrodes. It can be observed that it has an initial reversible discharge capacity of 950 mAh/g and its capacity keeps steady even after 100 cycles. In order to evaluate the charge–discharge efficiency, the coulombic efficiency of the electrodes was also exhibited in Figure 3c. The coulombic efficiency is defined as the percentage of discharge and charge capacity in one cycle. It is seen from Figure 3c that the coulombic efficiency is almost 100% after about 10 cycles. Therefore, it can be concluded that the ZnS nanotube electrodes have a good cycling stability, meanwhile delivering a quite high reversible capacity. This is due to the fact that the ZnS nanotubes possess a large specific area, which provides more active sites for the lithium ions [33]. That is, during the charge or discharge processes, tube shaped ZnS can accommodate more lithium ions [33]. Moreover, 1-D ZnS nanotubes can shorten the lithium ion diffusion distance [34]. At the same time, nanotube structure has a beneficial effect on the volume expansion/shrinkage of ZnS during charge–discharge process [35]. The above reasons are why the ZnS nanotube electrodes have a good rate discharge capability and long cycle life.

In order to further investigate the structural changes of the ZnS nanotubes during charge–discharging process, the phase composition and microstructure for the ZnS nanotubes after 100 charge–discharge cycles were performed by using XRD and SEM. As shown in Figure 3d, three diffraction peaks can be clearly observed, which corresponds to the (111), (220), and (311) crystal. However, the intensity of the diffraction peaks is smaller than that of the prepared ZnS nanotubes before electrochemical cycle (Figure 2a). This may be due to the lower testing sample quantity. In addition, the SEM image of the sample after 100 charge–discharge cycles shows that the ZnS nanotubes still maintain a tube shape structure, together with some small amount of broken tube walls. Clearly, the ZnS nanotubes have a good structural stability during charge–discharge cycling, which leads to a good discharge capacity retention ratio.

Figure 4a shows the CV curves of the ZnS nanotube electrode at a scan rate of 0.1 mV/s between 0.01 to 3 V for the first, second, and third discharge/charge processes. In the first discharge process, two broad reduction peaks can be observed in the potential range of 1.1–0.25 V and 0.25–0.05 V, respectively. The first reduction peak ranging from 1.1–0.25 V can be ascribed to the decomposition of ZnS into Zn and the formation of Li_2S, while the second reduction peak ranging from 0.25–0.05 V corresponds to the subsequent reaction of lithium ions with Zn metal [14]. Decomposition of the electrolyte and formation of the solid electrolyte interface on the surface of electrode particles also occurs during the first cathodic cycle, which causes part of the irreversible capacity [12,14]. During the anodic scanning process, anodic peaks located at 0.28, 0.37, 0.57, and 0.71 V are observed, which is attributed to the multi-step de-alloying process of Li-Zn (LiZn, Li_2Zn_3, $LiZn_2$, and Li_2Zn_5) [12]. Another bigger anodic peak at 1.26 V is attributed to the back-conversion of Zn and Li_2S into ZnS [12,14]. In the following cycles, the CV curves do not change position or intensity of any of the peaks, indicating that the ZnS nanotubes have a good reversibility.

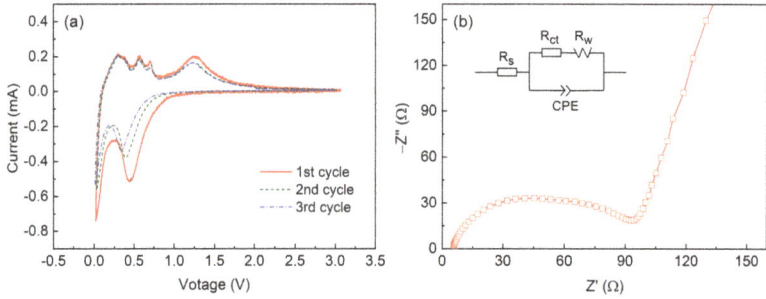

Figure 4. (**a**) CV curves of the ZnS nanotube electrode at a scan rate of 0.1 mV/s; (**b**) Nyquist plots of the ZnS electrodes.

To further confirm the electrochemical performance, the EIS studies of the ZnS nanotube electrodes were conducted, results are shown in Figure 4b. A semicircle and a slope line can be observed for both of the samples. As reported by Cheng et al. [36], the semicircles relate to the charge transfer resistance which resulted from charge transfer through the electrode/electrolyte interface, while the slope line corresponds to Warburg impedance over Li$^+$ diffusion in the solid materials. The impedance data were analyzed by fitting the curves with equivalent electrical circuit shown in the inset of Figure 4b, where R_s and R_{ct} indicate electrolyte resistance and charge transfer resistance, respectively; CPE stands for the corresponding constant phase elements; and R_W represents the diffusion-controlled Warburg impedance. The charge-transfer resistance of the as-synthesized ZnS nanotube electrodes is 88 Ω, which is much lower than the pure ZnS nanoparticles (199.5 Ω) reported by Zhang et al. [37]. Moreover, the ZnS nanotube electrode has a higher slope, which means the Li$^+$ diffusion in the ZnS nanotubes is much faster than that of ZnS nanoparticles. Clearly, ZnS nanotube electrodes exhibit a good electrochemical performance.

4. Conclusions

Comprehensive analysis—using XRD, SEM, TEM, and XPS—indicates that the ZnS nanotubes were successfully prepared via a hydrothermal method. The BET specific surface area of the sample is as high as 86.86 m^2/g. The ZnS nanotube electrodes exhibit a good lithium ion storage capability. The first discharge capacity was about 950 mAh/g, and it could still remain at 450 mAh/g after 100 charge/discharge cycles. This is due to the fact that the ZnS nanotubes have a good structural stability during charge–discharge cycling. In a word, this work provides a rational strategy to advance the ZnS anode electrochemical performance in lithium-ion batteries applications.

Author Contributions: Conceptualization, All authors; Methodology and experiment, W.Z., J.Z., and T.T.; Writing–Original Draft Preparation, W.Z. and J.Z.; Writing–Review & Editing, Y.Z. and T.Y.; Funding Acquisition, W.Z. and T.Y.

Funding: This research was funded by Hebei Provincial Natural Science Foundation China [E2018202140] and Youth Top Talents Research Project of Hebei Provincial Education Department China [BJ2018036].

Conflicts of Interest: The authors declare no conflict of interest.

References

1. Li, H.P.; Li, Y.; Zhang, Y.G.; Zhang, C.W. Facile synthesis of carbon-coated Fe$_3$O$_4$ core-shell nanoparticles as anode materials for lithium-ion batteries. *J. Nanopart. Res.* **2015**, *17*, 370. [CrossRef]
2. Liu, Z.H.; Chai, J.C.; Xu, G.J.; Wang, Q.F.; Cui, G.L. Functional lithium borate salts and their potential application in high performance lithium batteries. *Coord. Chem. Rev.* **2015**, *292*, 56–73. [CrossRef]
3. Han, J.H.; Liu, P.; Ito, Y.; Guo, X.W.; Hiratai, A.; Fujita, T.; Chen, M.W. Bilayered nanoporous graphene/molybdenum oxide for high rate lithium ion batteries. *Nano Energy* **2018**, *45*, 273–279. [CrossRef]
4. Xu, J.M.; Han, Z.; Wu, J.S.; Song, K.X.; Wu, J.; Gao, H.F.; Mi, Y.H. Synthesis and electrochemical performance of vertical carbon nanotubes on few-layer graphene as an anode material for Li-ion batteries. *Mater. Chem. Phys.* **2018**, *205*, 359–365. [CrossRef]
5. Ren, Y.R.; Wei, H.M.; Yang, B.; Wang, J.W.; Ding, J.N. "Double-Sandwich-Like" CuS@reduced graphene oxide as an anode in lithium ion batteries with enhanced electrochemical performance. *Electrochim. Acta* **2014**, *145*, 193–200. [CrossRef]
6. Zhang, Y.G.; Li, Y.; Li, H.P.; Yin, F.X.; Zhao, Y.; Bakenov, Z. Synthesis of hierarchical MoS$_2$ microspheres composed of nanosheets assembled via facile hydrothermal method as anode material for lithium-ion batteries. *J. Nanopart. Res.* **2016**, *18*, 1–9. [CrossRef]
7. Zhi, L.Z.; Han, E.S.; Cao, J.L. Synthesis and Characteristics of NiS for Cathode of Lithium Ion Batteries. *Adv. Mater. Res.* **2011**, *236*, 694–697. [CrossRef]
8. Mao, M.L.; Jiang, L.; Wu, L.C.; Zhang, M.; Wang, T.H. The structure control of ZnS/graphene composites and their excellent properties for lithium-ion battery. *J. Mater. Chem. A* **2015**, *3*, 13384–13389. [CrossRef]

9. Li, H.; Wang, Y.H.; Huang, J.X.; Zhang, Y.Y.; Zhao, J.B. Microwave-assisted synthesis of CuS/graphene composite for enhanced lithium storage properties. *Electrochim. Acta* **2017**, *225*, 443–451. [CrossRef]
10. Li, X.; Chen, Y.; Zou, J.; Zeng, X.; Zhou, L.; Huang, H. Stable freestanding Li-ion battery cathodes by in situ conformal coating of conducting polypyrrole on NiS-carbon nanofiber films. *J. Power Sources* **2016**, *331*, 360–365. [CrossRef]
11. Guo, J.; Chen, X.; Jin, S.; Zhang, M.; Liang, C. Synthesis of graphene-like MoS_2 nanowall/graphene nanosheethybrid materials with high lithium storage performance. *Catal. Today* **2015**, *246*, 165–171. [CrossRef]
12. Du, X.F.; Zhao, H.L.; Zhang, Z.J.; Lu, Y.; Gao, C.H.; Li, Z.L.; Teng, Y.Q.; Zhao, L.; Świerczek, K. Core-shell structured ZnS-C nanoparticles with enhanced electrochemical properties for high-performance lithium-ion batteries anodes. *Electrochim. Acta* **2017**, *225*, 129–136. [CrossRef]
13. Li, H.P.; Liu, Z.J.; Yang, S.; Zhao, Y.; Feng, Y.T.; Bakenov, Z.; Zhang, C.W.; Yin, F.X. Facile synthesis of ZnO nanoparticles on nitrogen-doped carbon nanotubes as high performance anode material for lithium-ion batteries. *Materials* **2017**, *10*, 1102. [CrossRef] [PubMed]
14. He, L.; Liao, X.Z.; Yang, K.; He, Y.S.; Wen, W.; Ma, Z.F. Electrochemical characteristics and intercalation mechanism of ZnS/C composite as anode active material for lithium-ion batteries. *Electrochim. Acta* **2011**, *56*, 1213–1218. [CrossRef]
15. Kim, I.; Kim, B.S.; Nam, S.; Lee, H.G.; Chung, H.K.; Cho, S.M.; Luu, T.H.T.; Hyun, S.; Kang, C. Cross-linked poly (vinylidene fluoride-co-hexafluoropropene) (PVDF-co-HFP) gel polymer electrolyte for flexible Li-ion battery integrated with organic light emitting diode (OLED). *Materials* **2018**, *11*, 543. [CrossRef] [PubMed]
16. Ma, J.J.; Wang, H.J.; Liu, X.; Liu, X.R.; Lu, L.D.; Nie, L.Y.; Yang, X.; Chai, Y.Q.; Yuan, R. Synthesis of tube shape MnO/C composite from 3, 4, 9, 10-perylenetetracarboxylic dianhydride for lithium ion batteries. *Chem. Eng. J.* **2017**, *309*, 545–551. [CrossRef]
17. Xu, G.L.; Xu, Y.F.; Sun, H.; Fu, F.; Zheng, X.M.; Huang, L.; Li, J.T.; Yang, S.H.; Sun, S.G. Facile synthesis of porous MnO/C nanotubes as a high capacity anode materialfor lithium ion batteries. *Chem. Commun.* **2012**, *48*, 8502–8504. [CrossRef] [PubMed]
18. Donne, A.L.; Jana, S.K.; Banerjee, S.; Basu, S.; Binetti, S. Optimized luminescence properties of Mn doped ZnS nanoparticles for photovoltaic applications. *J. Appl. Phys.* **2013**, *113*, 014903. [CrossRef]
19. Alvarez-Coronadoi, E.G.; González, L.A.; Rendón-Ángeles, J.C.; Meléndez-Lira, M.A.; Ramírez-Bon, R. Study of the structure and optical properties of Cu and Mn in situ doped ZnS films by chemical bath deposition. *Mater. Sci. Semicond. Process.* **2018**, *81*, 68–74. [CrossRef]
20. Liu, X.; Yang, Y.; Li, Q.; Wang, Z.Z.; Xing, X.X.; Wang, Y.D. Portably colorimetric paper sensor based on ZnS quantum dots for semi-quantitative detection of Co^{2+} through the measurement of grey level. *Sens. Actuators B Chem.* **2018**, *260*, 1068–1075. [CrossRef]
21. Reddy, A.L.M.; Shaijumon, M.M.; Gowda, S.R.; Ajayan, P.M. Coaxial MnO_2/carbon nanotube array electrodes for high-performance lithium batteries. *Nano Lett.* **2009**, *9*, 1002–1006. [CrossRef] [PubMed]
22. Yan, C.L.; Xue, D.F. Conversion of ZnO nanorod arrays into ZnO/ZnS nanocable and ZnS nanotube arrays via an in situ chemistry strategy. *J. Phys. Chem. B* **2006**, *110*, 25850–25855. [CrossRef] [PubMed]
23. Wang, Y.S.; Ma, Z.M.; Chen, Y.J.; Zou, M.C.; Yousaf, M.; Yang, Y.B.; Yang, L.S.; Cao, A.Y.; Han, R.P.S. Controlled synthesis of core–shell carbon@MoS_2 nanotube sponges as high-performance battery electrodes. *Adv. Mater.* **2016**, *28*, 10175–10181. [CrossRef] [PubMed]
24. Chen, J.J. Recent Progress in Advanced Materials for Lithium ion Batteries. *Materials* **2013**, *6*, 156–183. [CrossRef] [PubMed]
25. Pujari, R.B.; Lokhande, A.C.; Yadav, A.A.; Kim, J.H.; Lokhande, C.D. Synthesis of MnS microfibers for high performance flexible supercapacitors. *Mater. Des.* **2016**, *108*, 510–517. [CrossRef]
26. Ding, C.H.; Su, D.Z.; Ma, W.X.; Zhao, Y.J.; Yan, D.; Li, J.B.; Jin, H.B. Design of hierarchical CuS/graphene architectures with enhanced lithium storage capability. *Appl. Surf. Sci.* **2017**, *403*, 1–8. [CrossRef]
27. Wei, B.B.; Liang, H.F.; Wang, R.R.; Zhang, D.F.; Qi, Z.B.; Wang, Z.C. One-step synthesis of graphitic-C_3N_4/ZnS composites for enhanced supercapacitor performance. *J. Energy Chem.* **2018**, *27*, 472–477. [CrossRef]
28. Chang, Y.C. Complex ZnO/ZnS nanocable and nanotube arrays with high performance photocatalytic activity. *J. Alloy. Compd.* **2016**, *664*, 538–546. [CrossRef]

29. Park, A.R.; Jeon, K.J.; Park, C.M. Electrochemical mechanism of Li insertion/extraction in ZnS and ZnS/C anodes for Li-ion batteries. *Electrochim. Acta* **2018**, *265*, 107–114. [CrossRef]
30. Saha, S.; Sarkar, P. Electronic structure of ZnO/ZnS core/shell quantum dots. *Chem. Phys. Lett.* **2013**, *555*, 191–195. [CrossRef]
31. Kaplan, H.K.; Sarsıcı, S.; Akay, S.K.; Ahmetoglu, M. The characteristics of ZnS/Si heterojunction diode fabricated bythermionic vacuum arc. *J. Alloy. Compd.* **2017**, *724*, 543–548. [CrossRef]
32. Cui, L.F.; Hu, L.B.; Choi, J.W.; Cui, Y. Light-weight free-standing carbon nanotube-silicon films for anodes of lithium ion batteries. *ACS Nano* **2010**, *4*, 3671–3678. [CrossRef] [PubMed]
33. Renuga, V.; Mohan, C.N.; Manikandan, A. Influence of Mn^{2+} ions on both core/shell of $CuInS_2$/ZnS nanocrystals. *Mater. Res. Bull.* **2018**, *98*, 265–274. [CrossRef]
34. Qin, W.; Li, D.S.; Zhang, X.J.; Yan, D.; Hu, B.W.; Pan, L.K. ZnS nanoparticles embedded in reduced graphene oxide as high performance anode material of sodium-ion batteries. *Electrochim. Acta* **2016**, *191*, 435–443. [CrossRef]
35. Ortiz, G.F.; Hanzu, I.; Lavela, P.; Knauth, P.; Djenizian, T.; José, N.A.; Tirado, J. Novel fabrication technologies of 1D TiO_2 nanotubes vertical tin and iron-based nanowires for Li-ion microbatteries. *Int. J. Nanotechnol.* **2012**, *9*, 260–294. [CrossRef]
36. Cheng, S.Y.; Shi, T.L.; Tao, X.X.; Zhong, Y.; Huang, Y.Y.; Li, J.J.; Liao, G.L.; Tang, Z.R. In-situ oxidized copper-based hybrid film on carbon cloth as flexible anode for high performance lithium-ion batteries. *Electrochim. Acta* **2016**, *212*, 492–499. [CrossRef]
37. Zhang, R.P.; Wang, Y.; Jia, M.Q.; Xu, J.J.; Pan, E. One-pot hydrothermal synthesis of ZnS quantum dots/graphene hybrids as a dual anode for sodium ion and lithium ion batteries. *Appl. Surf. Sci.* **2018**, *437*, 375–383. [CrossRef]

 © 2018 by the authors. Licensee MDPI, Basel, Switzerland. This article is an open access article distributed under the terms and conditions of the Creative Commons Attribution (CC BY) license (http://creativecommons.org/licenses/by/4.0/).

Article

Enhanced Electrochemical Performance of Li$_{1.27}$Cr$_{0.2}$Mn$_{0.53}$O$_2$ Layered Cathode Materials via a Nanomilling-Assisted Solid-state Process

Chengkang Chang [1,2,*], **Jian Dong** [1], **Li Guan** [1] **and Dongyun Zhang** [1,*]

- [1] School of Materials Science and Engineering, Shanghai Institute of Technology,100 Haiquan Road, Shanghai 201418, China; theanswer0328@gmail.com (J.D.); lguan@sit.edu.cn (L.G.)
- [2] Shanghai Innovation Institute for Materials, Shanghai University, Shanghai 200444, China
- [*] Correspondence: ckchang@sit.edu.cn (C.C.); dyz@sit.edu.cn (D.Z.);
  Tel.: +86-135-8579-3649 (C.C.); +86-135-8579-3649 (D.Z.)

Received: 10 November 2018; Accepted: 27 January 2019; Published: 3 February 2019

Abstract: Li$_{1.27}$Cr$_{0.2}$Mn$_{0.53}$O$_2$ layered cathodic materials were prepared by a nanomilling-assisted solid-state process. Whole-pattern refinement of X-ray diffraction (XRD) data revealed that the samples are solid solutions with layered α-NaFeO$_2$ structure. SEM observation of the prepared powder displayed a mesoporous nature composed of tiny primary particles in nanoscale. X-ray photoelectron spectroscopy (XPS) studies on the cycled electrodes confirmed that triple-electron-process of the Cr^{3+}/Cr^{6+} redox pair, not the two-electron-process of Mn redox pair, dominants the electrochemical process within the cathode material. Capacity test for the sample revealed an initial discharge capacity of 195.2 mAh·g^{-1} at 0.1 C, with capacity retention of 95.1% after 100 cycles. EIS investigation suggested that the high Li ion diffusion coefficient (3.89 × 10^{-10}·cm^2·s^{-1}), caused by the mesoporous nature of the cathode powder, could be regarded as the important factor for the excellent performance of the Li$_{1.27}$Cr$_{0.2}$Mn$_{0.53}$O$_2$ layered material. The results demonstrated that the cathode material prepared by our approach is a good candidate for lithium-ion batteries.

Keywords: cathode material; X-ray diffraction; Cr^{3+}/Cr^{6+} redox pairs; specific capacity; cycling performance

1. Introduction

Manganese-based cathode materials have been widely studied for use in lithium-ion batteries due to their low cost, nontoxicity and, in the case of layered LiMnO$_2$, high theoretical capacity (285 mAh·g^{-1}). Advances in electrode materials are very important for the development of rechargeable lithium-ion batteries. As an electrode of a lithium-ion battery, several compounds such as spinel LiMn$_2$O$_4$, layered LiCoO$_2$, and LiNiO$_2$ have been extensively studied [1–3]. LiMnO$_2$ in the form of a layered compound having the structure R-3m is interesting as a cathode material, but suffers from severe capacity decay during cycling. Besides, layered LiMnO$_2$ transforms into a spinel structure during the lithium insertion/extraction process due to cation migration [4]. Recently, derivatives of layered manganese oxides, such as substituted LiM$_x$Mn$_{1-x}$O$_2$ (M = Al, Co, Cr, etc.) [5,6] and lithium-saturated solid solutions or nanocomposite Li$_2$MnO$_3$-LiMO$_2$ (M = Ni, Co, or Fe) [5–11], have been investigated, in order to obtain cathodes with prolonged structural integrities and enhanced electrochemical performance.

Recently, because of the high specific capacity of lithium-rich phase xLi$_2$MnO$_3$·(1-x)LiMO$_2$ (M = Cr, Co, Mn, Ni, or Fe) [11–18] composites, it has been widely studied for the use as cathode material in lithium secondary battery. Some research results have been reported on the layered structure LiCrO$_2$-Li$_2$MnO$_3$. It was reported that by a solution method and subsequent quenching method, Li[Cr$_x$Li$_{(1-x)/3}$Mn$_{2(1-x)/3}$]O$_2$ (0.1 ≤ x ≤ 0.4) with nanocomposite structure was synthesized by Park et al. [12]. The material exhibited a high capacity of 195 mA·g^{-1}, when the cutoff voltage was between 2.4V and 4.7 V and the current density was 11.98 mA·g^{-1}. Kim et al. [19] synthesized Li[Cr$_x$Li$_{(1/3-x/3)}$Mn$_{(2/3-2x/3)}$]O$_2$ (0 < x < 1) by the sol-gel method. At a specific current density of 5 mA·g^{-1}, Li[Cr$_x$Li$_{(1-x)/3}$Mn$_{2(1-x)/3}$]O$_2$ with x = 1/6 can exhibit a high reversible capacity of 230 mAh·g^{-1} when the voltage is between 2.0 V and 4.8 V. In addition, layered Li-Cr-Mn-O cathode materials related to the LiCrO$_2$-LiMnO$_2$-Li$_2$MnO$_3$ solid solution have been synthesized by the mixed hydroxide method [8,20]. The cathode material in terms of high capacity and stable cycling performance exhibits an average discharge capacity of 204 mAh·g^{-1} between 2.5 and 4.5 V versus Li/Li$^+$. It is reported by Wu et al. [20] that the LiMnO$_2$ cathode material can reduce the topographical change from orthogonal to monoclinic geometry while improving cycle performance and reversible capacity through a small amount of Cr doping. On the other hand, Ko et al. [21] reported the 0.55Li$_2$TiO$_3$-0.45LiCrO$_2$ composite have the highest initial discharge capacity of 203 mAh·g^{-1}, showing that the chromium ions can participate in the electrochemical reactions. However, almost all the materials mentioned above showed fast capacity fading during the cycling and new systems and synthesis method were required to prepare the material with high capacity and stable cycling performance.

In this work, layered Li$_{1.27}$Cr$_{0.2}$Mn$_{0.53}$O$_2$ powders were synthesized by a nanomilling-assisted solid-phase method. The obtained Li$_{1.27}$Cr$_{0.2}$Mn$_{0.53}$O$_2$ cathode delivered high capacity close to the theoretical value with good capacity retention of 95.1% after 100 cycles. Such good electrochemical behavior can be attributed to the mesoporous nature of the cathode particles which offer fast pathway for the migration of the Li ions. The layered Li$_{1.27}$Cr$_{0.2}$Mn$_{0.53}$O$_2$ material presented high potential as a candidate to meet the demands of LIBs with high energy density.

2. Experimental

2.1. Synthesis of Li$_{1.27}$Cr$_{0.2}$Mn$_{0.53}$O$_2$ Using the Solid-State Reaction Method

The Li$_{1.27}$Cr$_{0.2}$Mn$_{0.53}$O$_2$ cathode materials were prepared by solid reaction method. In a typical process, according to the chemical composition with certain molar ratio of Li:Cr:Mn = 1.32:0.2:0.53, lithium hydroxide, in a purity of 99%, was dissolved in distilled water. Then, Cr$_2$O$_3$ (99%, Aladdin, Shanghai, China) and MnO$_2$ (Aladdin, 99% pure) powders were added to the above solution to form a suspension with continuous stirring. The resulting suspension was poured into a ball mill and treated for 2 h to achieve an uniform slurry with particle size of 200–300 nm using Zirconia grits with a diameter of 0.4 mm as the grinding media at a speed of 2000 rpm. Next, the ball-milled slurry was spray dried and the precursor powders were obtained. Finally, the powders were calcined at 500 °C for 3 h and then fired at 950 °C for 12 h in nitrogen to form the target Li$_{1.27}$Cr$_{0.2}$Mn$_{0.53}$O$_2$ cathode materials.

2.2. Instrumentation

The phase purity and crystal structure determination of the prepared powders were identified by X-ray diffraction (XRD, TD3500, Tongda, Dandong, China) method with Cu Kα radiation (λ = 1.54056 Å) carried out at 40 kV and 30 mA. The data were collected in the 2θ range of 10 to 70°. Rietveld refinements of XRD patterns were carried out by using Jade 9 software (materials data Inc., Livermore, CA, USA). The morphologies of the samples were examined by scanning electron microscopy (SEM, Hitachi, SU8200, Tokyo, Japan). XPS measurements were carried out using an ESCALAB 250Xi spectroscopy (Thermo Fisher Scientific, Waltham, MA, American) with Al Kα radiation. The binding energy was calibrated with respect to the conductive C 1s (285.0 eV).

The electrochemical performances of the samples were investigated using electrodes CR2016 coin-type cells at 25 °C, which were assembled inside a glove box filled with Ar. In a typical process, a mixture of the calcined powders, carbon black and binder, in the appropriate weight ratio of (8:1:1) dispersed by N-methyl-2-pyrrolidone (NMP). The slurry was then cast on aluminum foil to form a sheet that was cut into a circle of 1.44 cm^2, and then dried in vacuum at 110 °C for 10 h. The loading of Active material was 4–5 mg per disk. Lithium metal foils were used as the working anode. The Celgard 3501 membrane (Kejing, Shengzhen, China) was employed as the separator. A special electrolyte provided by Dongguan Hangsheng (Dongguan, China), which can work at high voltage, was employed in the experiment. The electrolyte is composed of 1 M LiPF$_6$/fluoroethylene carbonate (FEC)−ethyl methyl carbonate (EMC) (3:7 in volume ratio), with 0.5 wt% LiDFOB additive [22]. The electrochemical performance of the prepared cathode materials was determined on a Land CT2001 battery tester (Lanhe, Wuhan, China) at the voltage of 2.0–4.9 V. The cyclic voltammetry (CV) studies were conducted at a scan rate of 0.0 5mV·s^{-1} with cut-off voltages of 2.0 V and 4.9 V. Electrochemical Impedance Spectroscopy (EIS) was conducted by an electrochemical workstation (Autolab Pgstat302n, Metrohm, Zofingen, Switzerland) and the data were collected in the frequency range of 0.05 to 500 KHz.

3. Result and Discussion

3.1. Phase Purity of Synthesized Cathode Powders

Figure 1a showed the XRD patterns of the cathode Li$_{1.27}$Cr$_{0.2}$Mn$_{0.53}$O$_2$ materials prepared at 900 °C, 950 °C, 1000 °C. The peaks can be indexed into a hexagonal α-NaFeO$_2$structure (space group R-3m), except several superlattice ordering peaks between 20° and 30° marked in the figure with arrows. These superlattices are regarded as an indicator of the coexistence of Li$_2$MnO$_3$ phase. In an earlier report by Kim et al. [23], the existence of Li$_2$MnO$_3$ phase will cause the formation of MnO$_3$ phase during the electrochemical process, which will release O$_2$ gas during a subsequent chemical process. The evolution of O$_2$ gas will reduce the stability of the cathode itself and lead to a fast decay in electrochemical performance. Therefore, the observation of superlattice in the XRD pattern can be regarded as an indicator to determine whether the sample is well prepared or not. By comparison, it is clear that the sample prepared at 950 °C is of high purity and therefore the synthesis temperature was fixed at 950 °C for the sample preparation thereafter. Furthermore, the splitting of the (006)/(012) and (018)/(110) can be observed from the pattern, indicating that the prepared cathode material has a highly ordered layered structure. The above XRD results strongly indicated the successful synthesis of a pure phase for the layered hexagonal product in the experiment, and the obtained Li$_{1.27}$Cr$_{0.2}$Mn$_{0.53}$O$_2$ compound can be regarded as a normal Li[Li$_{0.27}$Cr$_{0.2}$Mn$_{0.53}$]O$_2$ layered complex.

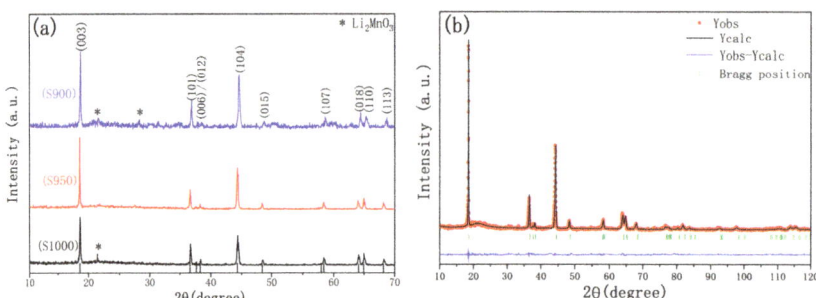

Figure 1. (a) X-ray diffractions of samples prepared. (b) Rietveld refinement results for sample S950.

Rietveld refinement for the sample S950 is presented in Figure 1b. The lattice parameters of the a axis and c axis were calculated using Jade 9. The lattice parameters and atomic occupancies of the sample was provided on the basis of high symmetry R-3m space group, as listed in Tables 1 and 2. Generally, the error indicators, R$_{wp}$ and R in Table 1, are two important factors for evaluating the refinement results, where Rwp is the weight distribution factor and R is the confidence factor, and it is reliable and acceptable when the R factor are below 10%. The observed and calculated patterns match well, so the refinement is acceptable. Furthermore, higher values of I_{003}/I_{104} listed in Table 2 correspond to the ideal layered structure of the cathode material. It is reported by Li et al. that when the I_{003}/I_{104} exceeded 1.311, no cationic disordering was presented [24]. In our approach, I_{003}/I_{104} value of 1.625 was obtained for sample S950, suggesting the absence of cationic disordering. Such a conclusion can also be obtained from the fact that Li1 (3b) sites were not filled with Cr/Mn cations in Table 1. In addition, more pronounced splitting of the (018) and (110) doublet greatly suggested that the cathode material is composed of a well-defined layered structure, and thus good electrochemical performance for the sample material was expected.

Table 1. Rietveld refinement results for the XRD patterns.

Sample	Atom	Wyckoff Position	x	y	z	Occupancy	R (%)	Rwp (%)
S950	Li (1)	3b	0	0	0.5	0.9998	9.49	10.44
	Li (2)	3a	0	0	0	0.2686		
	Mn (1)	3a	0	0	0	0.5274		
	Cr (1)	3a	0	0	0	0.1984		
	O (1)	6c	0	0	0.2416	1		

Table 2. Rietveld refinement results of lattice parameters for S950.

Sample	a(Å)	c(Å)	V(Å^3)	I_{003}/I_{104}
S950	2.8643	14.2645	101.35	1.625

3.2. Valence States of Mn and Cr Ions within the Cathode Powder

XPS investigations were conducted to confirm the valence states of the cathode powders. Since MnO_2 and Cr_2O_3 were employed as raw materials, Cr^{3+} and Mn^{4+} were expected to be present in the synthesized compound. The whole pattern survey for the $Li_{1.27}Cr_{0.2}Mn_{0.53}O_2$ powder is presented in Figure 2a, where peaks representing Cr, Mn, O, and C ions were observed and denoted on the figure. Detailed investigations of the Mn2p and Cr2p signals were shown in Figure 2b,c, where the spectra were deconvolved. For the spectrum of Mn2p in Figure 2b, both peaks can only be deconvolved into one single peak, suggesting the existence of Mn^{4+} ions. For the spectrum of Cr2p in Figure 2c, similar results were observed. Both the peaks forCr2p can only be only deconvolved into one single peak, indicating the presence of Cr^{3+} ions within the synthesized cathode powder. The fitting results were listed in Table 3. It can be found from Table 3 that the valence states of Cr and Mn ions in $Li_{1.27}Cr_{0.2}Mn_{0.53}O_2$ powder are +3 and +4, respectively. XPS results show that the valence of Mn and Cr ions are consistent with the valence state of the starting raw materials, as we expected.

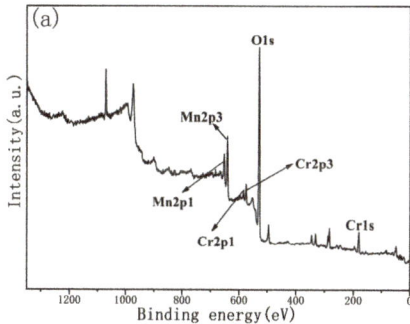

Figure 2. Cont.

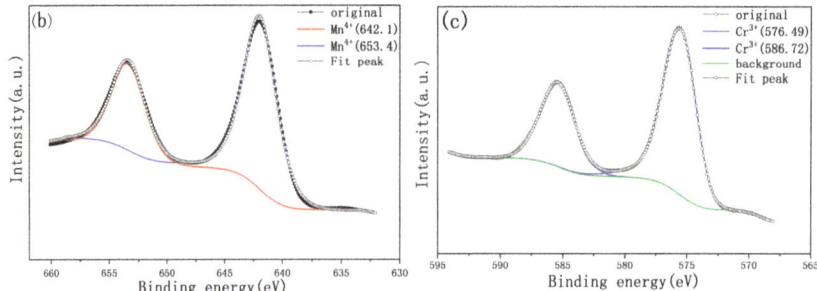

Figure 2. XPS spectra of the S950 powder (**a**). Whole pattern survey: (**b**) Mn2p and (**c**) Cr2p.

Table 3. XPS simulation results for $Li_{1.27}Cr_{0.2}Mn_{0.53}O_2$ powder sintered at 950 °C.

Source	Component	Valance State	BE/eV	FWHM/eV	Relative Area/%
Figure 2b	642.5	Mn^{4+}	642.1	1.9	100
		Mn^{3+}	-	-	0
Figure 2b	653.4	Mn^{4+}	653.37	2.1	100
		Mn^{3+}	-	-	0
Figure 2c	576.5	Cr^{3+}	576.49	1.96	100
		Cr^{6+}	-	-	0
Figure 2c	586.8	Cr^{3+}	586.72	2	100
		Cr^{6+}	-	-	0

3.3. SEM Observation of the Prepared Powder

The morphologies of the prepared samples were monitored by SEM. In general, the morphologies of the spray dried sample powders are characterized by uniform and large spherical agglomerates. Figure 3 shows the SEM images of $Li_{1.27}Cr_{0.2}Mn_{0.53}O_2$ prepared at 950 °C under N_2 atmosphere. It can be seen from Figure 3a that the powder is composed of uniform spherical particles with diameter of 2–4 µm. Figure 3b shows the enlarged micrograph of an individual particle, where tiny primary grains with size approximately 100–500 nm were observed. Microspores among the nanosized grains were also observed. N_2 adsorption/desorption test was conducted to get more details about the mesostructure. At the liquid nitrogen temperature, in the nitrogen-containing atmosphere, the surface of the powder will physically adsorb nitrogen. The specific surface area (S_{BET}) of the powder can be obtained by the following formula.

$$S_{BET} = 4.36V/W$$

where V is the adsorption amount of a complete layer of nitrogen molecules adsorbed on the surface of the powder. W is weight of sample. The adsorption/desorption plot and the size distribution of the nanopore within the mesostructure are presented in Figure 3c,d. The specific surface area (S_{BET}) can be calculated and the value was determined as 25.88 $m^2 \cdot g^{-1}$ for S950. An average diameter of 4–5 nm for the individual nanopore was also confirmed from Figure 3d. Due to the presence of this stable three-dimensional framework and mesoporous properties, the S950 sample will exhibit excellent cycle stability during charge–discharge. The numerous nanoparticles which aggregate to form porous microspheres can also enhance the transmission of Li ions by providing a short path for the intercalation/deintercalation of lithium-ions. It is possible to observe mesopores interconnected in the primary particles, which is advantageous for increasing the transport of Li^+ in the microspheres through the liquid electrolyte. Such nano-micro structures of the spherical cathode powders provide both short migration pathway and improved surface area for the redox process and thus good electrochemical performance could be expected.

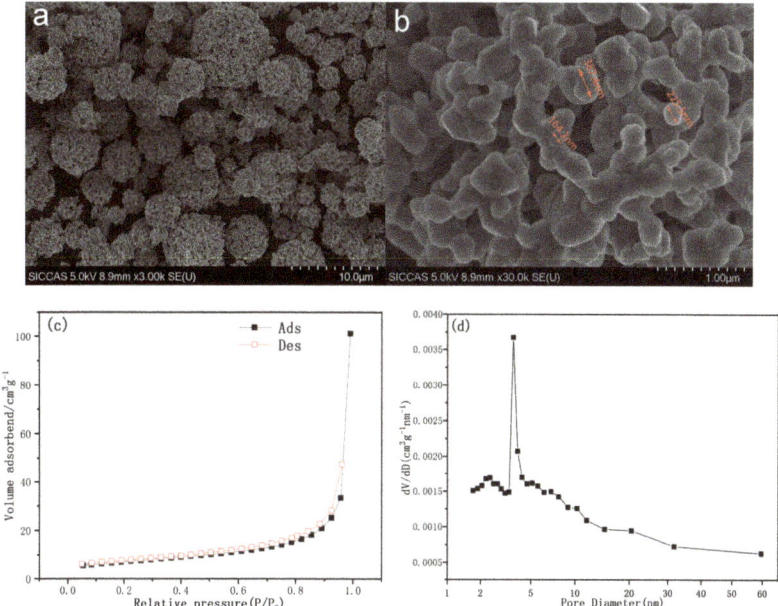

Figure 3. (**a,b**) SEM powder overview and individual particle, (**c**) nitrogen adsorption/desorption isotherms, and (**d**) mesopore distribution of $Li_{1.27}Cr_{0.2}Mn_{0.53}O_2$.

3.4. Electrochemical Performance

The electrochemical performances of the samples prepared at different temperatures were compared. The charge–discharge curves at the first cycle are presented and compared in Figure 4. It can be seen from the charging curves that, all three charging curves showed a gradually declined tendency, no obvious charging plateau was observed. Such charging curves greatly suggested the solid solution transition during the electrochemical process, which is very similar to the behavior of the other families of $LiMO_2$ layered structures [25]. Furthermore, it is also clear from the figure that the powder fired at 950 °C offered a highest specific capacity of 195.2 mAh·g^{-1}. According to the formula of $Li_{1.27}Cr_{0.2}Mn_{0.53}O_2$, if 0.2 mole Cr^{3+}/Cr^{6+} redox pairs were employed in the electrochemical process, 0.6 mole electrons will be released, and the nominal cathode compound will present a theoretical capacity of 200.18 mAh·g^{-1}. Therefore, in our experiment, by using Cr_2O_3 and MnO_2 as the starting materials, electrochemical cycles using Cr^{3+}/Cr^{6+} redox pairs, rather than the Mn redox pairs, were achieved with specific capacity close to its theoretical value.

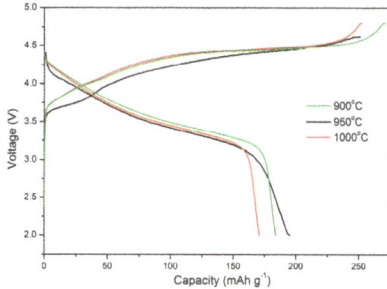

Figure 4. The initial charge–discharge curves of samples sintered with different temperature.

To confirm the solid solution transition manner of the electrochemical process, XRD investigations of the electrodes at charging and discharging states were compared to the pristine electrode before the capacity test, as shown in Figure 5. It is clear from Figure 5a that similar patterns were obtained for electrode samples at different state of charging. The pattern for the pristine electrode before cycling (pattern a1) showed good agreement to the XRD pattern of the $Li_{1.27}Cr_{0.2}Mn_{0.53}O_2$ powder sample, indicating the hexagonal nature of the cathode material before electrochemical cycling. Pattern a2 and pattern a3 showed the XRD results after the first charging/discharging process. All three patterns looked very alike, implying the same crystal structure of hexagonal type throughout the electrochemical process. Figure 5b showed an enlarged pattern at 2θ range of 43° to 46°. It is clear from the figure that no diffraction peak splitting was observed, only slight peak position shift was presented. Such results confirmed the solid solution transition manner of the cathode material during the Li intercalation and deintercalation process, rather than the two-phase transition manner, which is usually judged by the peak splitting.

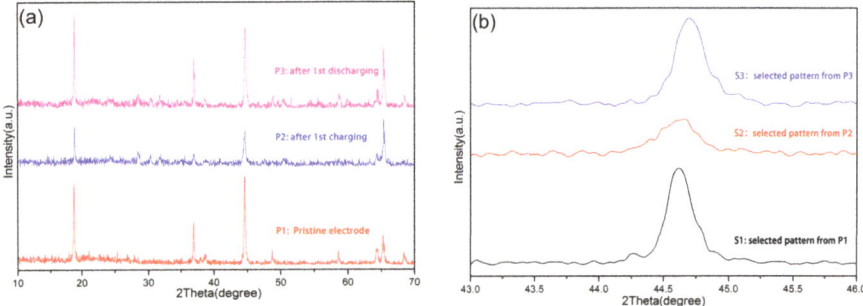

Figure 5. XRD for the electrodes: (**a**) whole range pattern and (**b**) at selected range.

XPS investigations of the electrodes at different SOC were conducted to further confirm the valence of chromium ions and manganese ions during the cycling. Figure 6 shows the changes in the valence state of Cr and Mn ions during charge and discharge processes. Detailed investigations of the Mn2p signals were shown in Figures 6a and 6c, where the spectra were deconvolved. Figure 6a,c shows two main peaks in the Mn2p spectra, which can be assigned to the manganese $2p_{3/2}$ at 641.8 eV and $2p_{1/2}$ at 654.0 eV. The binding energy is well consistent with the data on MnO_2, meaning Mn^{4+} in the $Li_{1.27}Cr_{0.2}Mn_{0.53}O_2$. Therefore, as can be seen from Figure 6a,c, the Mn ions remain at +4 valance states and do not participate in the electrochemical reactions during charging and discharging process. However, for Cr2p signals, the peaks can be deconvolved into two peaks, indicating the presence of Cr^{3+} ions after charging and Cr^{6+} ions after discharging, as can be seen from the curves in Figure 6b,d. Therefore, by comparison, it can be deduced that Cr^{3+}/Cr^{6+} redox pairs played the dominant role in the electrochemical process.

The fitting results from software Avantage were listed in Table 4. It is quite clear from Table 4 the Mn ions were kept very stable as +4 in valance state for both samples regardless charging or discharging state and no other valance state for Mn ions were detected. However, the difference in the valence state of Cr ions was obvious. For the sample after charging, most Cr ions were kept as +6, while it were kept as +3 for sample after discharging. Only a small amount of Cr^{3+} ions (~4 atm%) were observed in the charged electrode and a small portion of Cr^{6+} ions (~2.3 atm%) were detected in the discharged electrode. Therefore, from the above result, it is clear that Cr^{3+}/Cr^{6+} redox pairs, rather than the Mn^{4+}/Mn^{3+} redox pairs, decided the electrochemical performance of the prepared cathode material.

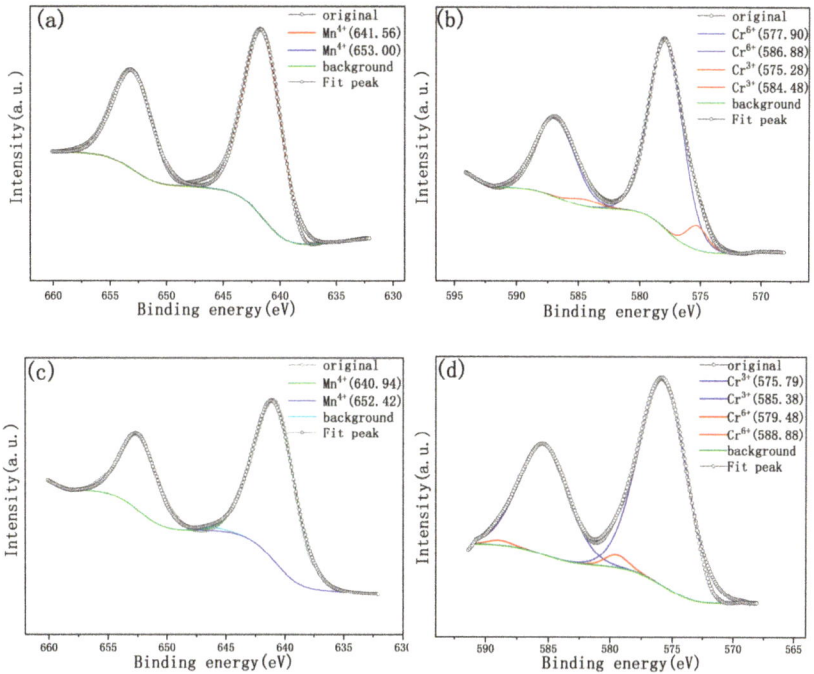

Figure 6. Deconvolved XPS spectra of electrodes at different SOC. (**a**) Mn2p at charging state, (**b**) Cr2p at charging state, (**c**) Mn2p at discharging state, and (**d**) Cr2p at discharging state.

Table 4. X-ray photoelectron spectroscopy (XPS) simulation results for the electrode.

Source	Component/eV	Valance State	BE/eV	FWHM/eV	Relative Area/%
Figure 6a Charged electrode	641.55	Mn^{4+}	641.56	1.71	100
		Mn^{3+}	-	-	0
Figure 6a Charged electrode	653.00	Mn^{4+}	653.03	1.55	100
		Mn^{3+}	-	-	0
Figure 6b Charged electrode	578.01	Cr^{6+}	577.85	1.73	95.8
		Cr^{3+}	575.19	1.08	4.2
Figure 6b Charged electrode	587.43	Cr^{6+}	586.97	1.93	96.2
		Cr^{3+}	584.68	1.49	3.8
Figure 6c Discharged electrode	641.01	Mn^{4+}	640.94	1.34	100
		Mn^{3+}	-	-	0
Figure 6c Discharged electrode	652.42	Mn^{4+}	652.45	1.76	100
		Mn^{3+}	-	-	0
Figure 6d Discharged electrode	576.21	Cr^{6+}	579.48	0.58	2.5
		Cr^{3+}	575.79	1.75	97.5
Figure 6d Discharged electrode	585.7	Cr^{6+}	588.88	0.53	2.1
		Cr^{3+}	585.38	1.79	97.9

The voltammetric behavior of the $Li_{1.27}Cr_{0.2}Mn_{0.53}O_2$ electrodes after 1, 2, and 50 cycles in the voltage range of 2.5 to 5.0 V are presented in Figure 7. It is clear that all the three curves are very similar, and the noticeable anodic peak around 3.6 V and cathodic peak around 3.2 V for the first cycle were observed. For the Cr^{3+}/Cr^{6+} redox couple in other system, such as $NaCrO_2$, the anodic scan showed a sharp and intense peak at 3.03 V and a low intensity peak at 3.3 V [26]. The corresponding cathodic peaks are observed at 2.8 V and 3.25 V, respectively. These peaks were associated with

Cr^{3+}/Cr^{4+} redox couple besides the Cr^{3+}/Cr^{6+} redox couple. In some other compounds, the cyclic voltammograms of $LiCr_{0.05}Ni_{0.45}Mn_{0.5}O_2$ and $LiCr_{0.1}Ni_{0.4}Mn_{0.5}O_2$ [10] revealed oxidation peaks in 2.9, 3.9, and 4.4 V regions and reduction peaks in the 2.7 and 3.6 V regions; additionally, more complicated phase transformation was observed. The same happened in our case, an oxidation peak at 4.5 V and a reduction peak at 4.13 V were also observed in the CV profile, which indicates that Cr ions in other valance state could be involved in the electrochemical process. The coexistence of multivalance state of Cr ions implies the complex of the electrochemical process, in which the transitions between $Cr^{3+}/Cr^{4+}/Cr^{6+}$ could take place during charging and discharging processes.

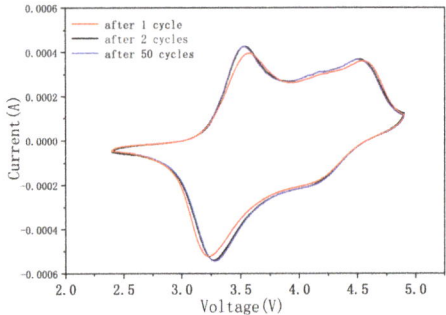

Figure 7. Cyclic voltammetry (CV) for the $Li_{1.27}Cr_{0.2}Mn_{0.53}O_2$ cell obtained at a scan rate of 0.05 mV·s^{-1}.

Furthermore, it can be found form the figure that at the second run, the anodic peak shifts to the negative potential of 3.4 V, while the cathodic peak shifted to the potential of 3.3 V, a very small $\Delta V(0.1V)$ was observed for the prepared $Li_{1.27}Cr_{0.2}Mn_{0.53}O_2$ electrode. For the runs thereafter until the 50th run, the redox pair at 3.3/3.4 V kept unchanged. This reinforces the conclusion that the material does not transform into other structures (like spinel in other layered phase) during cycling [18,27,28]. Such electrochemical behavior indicates that stable cycling has been set up within the electrode and therefore high performance of the electrodes can be expected.

Figure 8a shows the charge–discharge profiles with the cutoff voltages of 2.0 and 4.9 V at a rate of 0.1 C. The initial discharge specific capacity of $Li_{1.27}Cr_{0.2}Mn_{0.53}O_2$ was found to be 195.2 mAh·g^{-1} at the first cycle. It increased gradually onto 205.4 mAh·g^{-1} at the 10th run and then slightly dropped down at subsequent runs. With further cycling, both the capacity and the voltage profile became stable, which indicated that the structure is stable over the entire intercalation–deintercalation process. In addition, the degree of polarization gradually decreases with the increased cycles and it became relatively stable until the 10th cycle. The above results greatly implied that the $Li_{1.27}Cr_{0.2}Mn_{0.53}O_2$ cathode material presents excellent cycling stability at room temperature, regardless the difference in the voltage profiles. On the other hand, no plateau was observed at ~2.9 V on the profile, which indicates that the prepared material does not convert into a spinel phase, which was reported by other groups [18]. Figure 8b shows the cyclic performance of the $Li_{1.27}Cr_{0.2}Mn_{0.53}O_2$ cathode material. The discharging capacity increased up to 205.4 mAh·g^{-1} in the 10th cycle and then was maintained almost constant beyond the subsequent cycles. Moreover, it can be seen from Figure 8b that the capacity retention was 95.1% during the 100th cycles. All the results suggested an outstanding electrochemical performance of the prepared cathode.

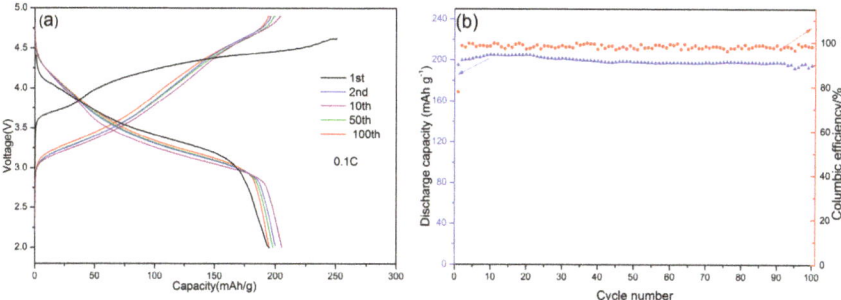

Figure 8. Charge–discharge curves and columbic efficiency of $Li_{1.27}Cr_{0.2}Mn_{0.53}O_2$ (**a,b**).

The electrochemical impedance spectra of $Li_{1.27}Cr_{0.2}Mn_{0.53}O_2$ measured after 1, 30, and 100 cycles are shown in Figure 9 in an attempt to figure out the reason for the high performance of the material. The resistivity of the sample is represented by the intercept of the semicircle and the real axis (Z'). In the Nyquist representation, due to the interfacial migration of lithium-ions between the surface layer and the electrolyte, the impedance data will appear in a semicircular form in the high-frequency region. [29]. The semicircle reduced significantly from 1st to 100th cycle of discharge. Wu et al. indicates that the reduction of charge transfer resistance can effectively accelerate the kinetics of the intercalation–deintercalation process, which is shown by a significant reduction in the semicircle in the figure [18]. Such impedance data can be simulated with an equivalent circuit and the solution resistance and the charge transfer resistance could be obtained. A suitable equivalent circuit containing Rs, Rct, CPE, and Zw showed the best fitting results for the experimental data, and the output results for Rs and Rct are listed in Table 5. For the value of Rs, a recording of only several ohms was obtained with a very slight change, since it represents for the resistance of the electrolyte itself. However, in term of Rct, the value varies from 233.88 Ω to 91.86 Ω, which indicates an obvious decrease in interfacial resistance. Hence, in our case, high specific capacity with stable columbic efficiency for the sample S950 was obtained during the cycling.

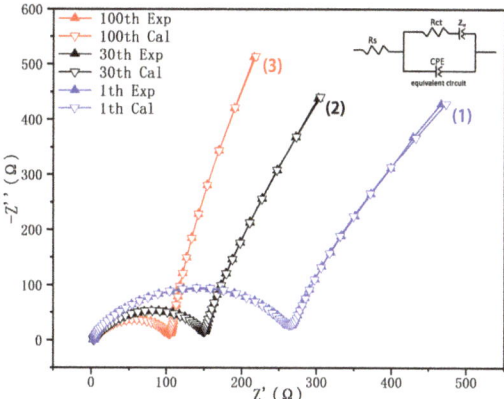

Figure 9. The impedance spectra for the $Li_{1.27}Cr_{0.2}Mn_{0.53}O_2$ electrode: (1) after first cycle, (2) after 30th cycle, and (3) after 100th cycle.

Table 5. Electrochemical impedance spectroscopy (EIS): initial resistance (Rs) and charge transfer resistance (Rct) after cycling.

Cycles	Rs(Ω)	Rct(Ω)
1th cycle	3.50	233.88
30th cycle	2.64	132.88
100th cycle	3.34	91.86

In addition, the following equation expresses the calculation equation of the diffusion coefficient of lithium-ions (D_{Li^+}) [30,31].

$$D_{Li^+} = \frac{0.5R^2T^2}{A^2F^4C^2\sigma_W^2} \quad (1)$$

where A the surface area of the electrode, T the absolute temperature, F the Faraday constant, R is the gas constant, C the concentration of Li$^+$ in the material, and σ_w is the Warburg factor obeying the following relationship,

$$Z_{Re} = R_S + R_{ct} + \sigma\omega^{-0.5} \quad (2)$$

σ_w can be obtained by fitting of the real parts of impedance Z_{Re} vs. $\omega^{-0.5}$ in the frequency range of 0.05 Hz to 1 Hz as shown in the Figure 10. As a result, the diffusion coefficients of the Li$_{1.27}$Cr$_{0.2}$Mn$_{0.53}$O$_2$ material calcined at 900 °C (1.05 × 10^{-10} cm^2·s^{-1}), 950 °C (3.89 × 10^{-10} cm^2·s^{-1}), and 1000°C (1.34 × 10^{-10} cm^2·s^{-1}) were determined. Furthermore, by calculation, the diffusion coefficient of the material S950 after 100 cycles (3.89 × 10^{-10} cm^2·s^{-1}) is higher than the values of Li$_{1.2}$Ni$_{0.2}$Mn$_{0.2}$O$_2$ (1.63 × 10^{-12} cm^2·s^{-1}) and Li$_{1.23}$Mn$_{0.46}$Ni$_{0.15}$Co$_{0.16}$O$_2$ (2.78 × 10^{-15} cm^2·s^{-1}) [31] reported by others. Such a high diffusion coefficient are supposed to provide a fast Li diffusion within the layered Li$_{1.27}$Cr$_{0.2}$Mn$_{0.53}$O$_2$ based on the Cr^{3+}/Cr^{6+} redox repairs, and high potential of the material as a cathode for LIBs could be achieved.

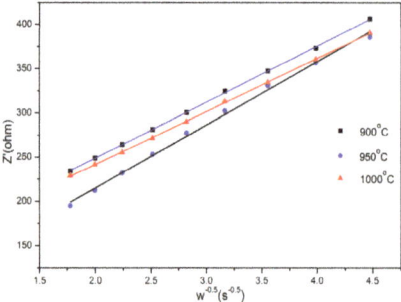

Figure 10. Plots comparison of Z' vs. $\omega^{-0.5}$ for Li$_{1.27}$Cr$_{0.2}$Mn$_{0.53}$O$_2$ after 100 cycles.

From the above results, it can be found that the sample S950 presented the best electrochemical performance among the three samples. The S950 cathode delivered an initial discharge capacity of 195.2 mAh·g^{-1} with capacity retention of 95.1% after 100 cycles. All the results can be explained by the fact that a fast Li ion transportation was achieved since the largest Li ion diffusion coefficient was observed for the sample.

4. Conclusions

Li$_{1.27}$Cr$_{0.2}$Mn$_{0.53}$O$_2$ cathode materials with hexagonal α-NaFeO$_2$ structure (space group R-3m) were successfully prepared by a nanomilling-assisted solid-state reaction method. The initial XPS studies confirmed the existence of Cr^{3+} and Mn^{4+} in the prepared powders as expected. The Li$_{1.27}$Cr$_{0.2}$Mn$_{0.53}$O$_2$ electrode showed a typical solid solution transition during the electrochemical cycling and no other side process was observed. CV tests revealed an oxidation/reduction couple of 3.6/3.2 V ascribed to the Cr^{3+}/Cr^{6+} redox pair rather than the Mn redox pairs. XPS studies on the electrodes after charging/discharging further confirmed that the triple-electron process of the Cr^{3+}/Cr^{6+} redox pair, not the Mn redox pairs, was superior to the electrochemical process within the cathode material. The capacity test for the sample exposed a high reversible capacity of ~195.2 mAh·g^{-1} (close to its theoretical value) between 2.0 and 4.9 V at 0.1 C, with capacity retention of 95.1% after 100 cycles. EIS investigation revealed a high diffusion coefficient of 3.89×10^{-10} cm^2·s^{-1}, which can be regarded as the reason for the good electrochemical performance of the prepared cathode material. As a whole, the prepared compound presented high potential as a candidate for the use as cathode in LIBs.

Author Contributions: C.C. and D.Z. designed the experimental process; L.G. and J.D. performed the experiments. J.D., L.G., D.Z., and C.C. analyzed the data and wrote the manuscript.

Funding: The research was supported by the Science and Technology Commission of Shanghai Municipality (14520503100 and 201310-JD-B2-009) and the Shanghai Municipal Education Commission (15ZZ095).

Conflicts of Interest: The authors declare no conflicts of interest.

References

1. Yu, Z.M.; Zhao, L.C. Structure and electrochemical properties of LiMn$_2$O$_4$. *Trans. Nonferrous Met. Soc. China* **2007**, *17*, 659–664. [CrossRef]
2. Aziz, N.A.A.; Abdullah, T.K.; Mohamad, A.A. Synthesis of LiCoO$_2$ Prepared by Sol-gel Method. *Procedia Chem.* **2016**, *19*, 861–864. [CrossRef]
3. Cheng, E.J.; Taylor, N.J.; Wolfenstine, J.; Sakamoto, J. Elastic properties of lithium cobalt oxide (LiCoO$_2$). *J. Asian Ceram. Soc.* **2017**, *5*, 113–117. [CrossRef]
4. Zheng, Y.; Hao, X.; Niu, J.; Pan, B. Layered o-LiMnO$_2$ prepared for lithium ion batteries by mechanical alloying. *Mater. Lett.* **2016**, *163*, 98–101. [CrossRef]
5. Wen, J.G.; Bareño, J.; Lei, C.H.; Kang, S.H.; Balasubramanian, M.; Petrov, I.; Abraham, D.P. Analytical electron microscopy of Li$_{1.2}$Co$_{0.4}$Mn$_{0.4}$O$_2$ for lithium-ion batteries. *Solid State Ionics* **2011**, *182*, 98–107. [CrossRef]
6. Zhang, X.; Meng, X.; Elam, J.W.; Belharouak, I. Electrochemical characterization of voltage fade of Li$_{1.2}$Ni$_{0.2}$Mn$_{0.6}$O$_2$ cathode. *Solid State Ionics* **2014**, *268*, 231–235. [CrossRef]
7. Li, J.; Jeong, S.; Kloepsch, R.; Winter, M.; Passerini, S. Improved electrochemical performance of LiMO$_2$ (M = Mn, Ni, Co)–Li$_2$MnO$_3$ cathode materials in ionic liquid-based electrolyte. *J. Power Sources* **2013**, *239*, 490–495. [CrossRef]
8. Kim, K.S.; Lee, S.W.; Moon, H.S.; Kim, H.J.; Cho, B.W.; Cho, W.I.; Choi, J.B.; Park, J.W. Electrochemical properties of Li–Cr–Mn–O cathode materials for lithium secondary batteries. *J. Power Sources* **2004**, *129*, 319–323. [CrossRef]
9. Zhang, B.; Li, L.; Zheng, J. Characterization of multiple metals (Cr, Mg) substituted LiNi$_{0.8}$Co$_{0.1}$Mn$_{0.1}$O$_2$ cathode materials for lithium ion battery. *Int. J. Min. Met. Mater.* **2012**, *520*, 190–194. [CrossRef]
10. Kim, G.T.; Kim, J.U.; Sim, Y.J.; Kim, K.W. Electrochemical properties of LiCr$_x$Ni$_{0.5-x}$Mn$_{0.5}$O$_2$ prepared by co-precipitation method for lithium secondary batteries. *J. Power Sources* **2006**, *158*, 1414–1418. [CrossRef]
11. Ko, Y.N.; Kim, J.H.; Lee, J.K.; Kang, Y.C.; Lee, J.H. Electrochemical properties of nanosized LiCrO$_2$·Li$_2$MnO$_3$ composite powders prepared by a new concept spray pyrolysis. *Electrochim. Acta* **2012**, *69*, 345–350. [CrossRef]
12. Wu, X.; Ryu, K.S.; Hong, Y.S.; Park, Y.J.; Chang, S.H. Properties of Li[Cr$_x$Li$_{(1-x)/3}$Mn$_{2(1-x)/3}$]O$_2$ ($0.1 \leq x \leq 0.2$) material prepared by quenching. *J. Power Sources* **2004**, *132*, 219–224. [CrossRef]
13. Park, C.W.; Kim, S.H.; Nahm, K.S.; Chung, H.T.; Lee, Y.S.; Lee, J.H.; Boo, S.; Kim, J. Structural and electrochemical study of Li[Cr$_x$Li$_{(1-x)/3}$Mn$_{2(1-x)/3}$]O$_2$ ($0 \leq x \leq 0.328$) cathode materials. *J. Alloys Compod.* **2008**, *449*, 343–348. [CrossRef]
14. Park, C.W.; Kim, S.H.; Mangani, I.R.; Lee, J.H.; Boo, S.; Kim, J. Synthesis and materials characterization of Li$_2$MnO$_3$–LiCrO$_2$ system nanocomposite electrode materials. *Mater. Res. Bull.* **2007**, *42*, 1374–1383. [CrossRef]
15. Yang, J.; Guo, B.; He, H.; Li, Y.; Song, C.; Liu, G. LiNi$_{0.5}$Mn$_{0.5}$O$_2$ hierarchical nanorods as high-capacity cathode materials for Li-ion batteries. *J. Alloys Compod.* **2017**, *698*, 714–718. [CrossRef]

16. Sun, Y.; Shiosaki, Y.; Xia, Y.; Noguchi, H. The preparation and electrochemical performance of solid solutions LiCoO$_2$–Li$_2$MnO$_3$ as cathode materials for lithium ion batteries. *J. Power Sources* **2006**, *159*, 1353–1359. [CrossRef]
17. Liao, D.Q.; Xia, C.Y.; Xi, X.M.; Zhou, C.X.; Xiao, K.S.; Chen, X.Q.; Qin, S.B. Li Rich Layered Cathode Material Li[Li$_{0.157}$Ni$_{0.138}$Co$_{0.134}$Mn$_{0.571}$]O$_2$ Synthesized with Solid-State Coordination Method. *J. Electron. Mater.* **2016**, *45*, 2981–2986. [CrossRef]
18. Wu, X.; Chang, S.H.; Park, Y.J.; Ryu, K.S. Studies on capacity increase of Li$_{1.27}$Cr$_{0.2}$Mn$_{0.53}$O$_2$-based lithium batteries. *J. Power Sources* **2004**, *137*, 105–110. [CrossRef]
19. Kim, S.; Noh, J.K.; Yu, S.; Chang, W.; Chung, K.Y.; Cho, B.W. Effects of transition metal doping and surface treatment to improve the electrochemical performance of Li$_2$MnO$_3$. *J. Electroceram.* **2013**, *30*, 159–165. [CrossRef]
20. Pang, W.K.; Lee, J.Y.; Wei, Y.S.; Wu, S.H. Preparation and characterization of Cr-doped LiMnO$_2$ cathode materials by Pechini's method for lithium ion batteries. *Mater. Chem. Phys.* **2013**, *139*, 241–246. [CrossRef]
21. Ko, Y.N.; Choi, S.H.; Kang, Y.C.; Park, S.B. Characteristics of Li$_2$TiO$_3$–LiCrO$_2$ composite cathode powders prepared by ultrasonic spray pyrolysis. *J. Power Sources* **2013**, *244*, 336–343. [CrossRef]
22. Xia, L.; Lee, S.; Jiang, Y.; Xia, Y.; Chen, G.Z.; Liu, Z. Fluorinated electrolytes for li-ion batteries: The lithium difluoro (oxalato) borate additive for stabilizing the solid electrolyte interphase. *ACS Omega* **2017**, *2*, 8741–8750. [CrossRef]
23. Kim, J.S.; Johnson, C.S.; Vaughey, J.T.; Thackeray, M.M.; Hackney, S.A.; Yoon, W.; Grey, C.P. Electrochemical and Structural Properties of x Li$_2$M'O$_3$·(1 − x) LiMn$_{0.5}$Ni$_{0.5}$O$_2$ Electrodes for Lithium Batteries (M' = Ti, Mn, Zr; $0 \leq x \leq 0.3$). *Chem. Mater.* **2004**, *16*, 1996–2006. [CrossRef]
24. Li, D.; Sasaki, Y.; Kobayakawa, K.; Sato, Y. Morphological, structural, and electrochemical characteristics of LiNi$_{0.5}$Mn$_{0.4}$M$_{0.1}$O$_2$ (M = Li, Mg, Co, Al). *J. Power Sources* **2006**, *157*, 488–493. [CrossRef]
25. Ganter, M.J.; DiLeo, R.A.; Schauerman, C.M.; Rogers, R.E.; Raffaelle, R.P.; Landi, B.J. Differential scanning calorimetry analysis of an enhanced LiNi$_{0.8}$Co$_{0.2}$O$_2$ cathode with single wall carbon nanotube conductive additives. *Electrochim. Acta* **2011**, *56*, 7272–7277. [CrossRef]
26. Ding, J.J.; Zhou, Y.N.; Sun, Q.; Fu, Z.W. Cycle performance improvement of NaCrO$_2$ cathode by carbon coating for sodium ion batteries. *Electrochem. Commun.* **2012**, *22*, 85–88. [CrossRef]
27. Wu, X.; Kim, S.B. Improvement of electrochemical properties of LiNi$_{0.5}$Mn$_{1.5}$O$_4$ spinel. *J. Power Sources* **2002**, *109*, 53–57. [CrossRef]
28. Shaju, K.M.; Subba Rao, G.V.; Chowdari, B.V.R. Electrochemical Kinetic Studies of Li-Ion in O$_2$-Structured Li$_{2/3}$(Ni$_{1/3}$Mn$_{2/3}$)O$_2$ and Li$_{(2/3)+x}$(Ni$_{1/3}$Mn$_{2/3}$)O$_2$ by EIS and GITT. *J. Electrochem. Soc.* **2003**, *150*, A1–A13. [CrossRef]
29. Levi, M.D. Solid-State Electrochemical Kinetics of Li-Ion Intercalation into Li$_{1-x}$CoO$_2$: Simultaneous Application of Electroanalytical Techniques SSCV, PITT, and EIS. *J. Electrochem. Soc.* **1999**, *146*, 1279. [CrossRef]
30. Li, X.; Xin, H.; Liu, Y.; Li, D.; Yuan, X.; Qin, X. Effect of niobium doping on the microstructure and electrochemical properties of lithium-rich layered Li[Li$_{0.2}$Ni$_{0.2}$Mn$_{0.6}$]O$_2$ as cathode materials for lithium ion batteries. *RSC Adv.* **2015**, *5*, 45351–45358. [CrossRef]
31. Luo, D.; Fang, S.; Yang, L.; Hirano, S. Improving the electrochemical performance of layered Li-rich transition-metal oxides by alleviating the blockade effect of surface lithium. *J. Mater. Chem. A* **2016**, *4*, 5184–5190. [CrossRef]

© 2019 by the authors. Licensee MDPI, Basel, Switzerland. This article is an open access article distributed under the terms and conditions of the Creative Commons Attribution (CC BY) license (http://creativecommons.org/licenses/by/4.0/).

Article

Effect of Different Composition on Voltage Attenuation of Li-Rich Cathode Material for Lithium-Ion Batteries

Jun Liu [1], Qiming Liu [1], Huali Zhu [2], Feng Lin [1], Yan Ji [1], Bingjing Li [1], Junfei Duan [1], Lingjun Li [1] and Zhaoyong Chen [1],*

1. College of Materials Science and Engineering, Changsha University of Science and Technology, Changsha 410114, China; liujun@stu.csust.edu.cn (J.L.); liuqiming@stu.csust.edu.cn (Q.L.); 18216359528@163.com (F.L.); juefly@stu.csust.edu.cn (Y.J.); krystalbingjingli@163.com (B.L.); junfei_duan@csust.edu.cn (J.D.); lingjun.li@csust.edu.cn (L.L.)
2. College of Physics and Electronic Science, Changsha University of Science and Technology, Changsha 410114, China; juliezhu2005@126.com
* Correspondence: chenzhaoyongcioc@126.com

Received: 30 November 2019; Accepted: 18 December 2019; Published: 20 December 2019

Abstract: Li-rich layered oxide cathode materials have become one of the most promising cathode materials for high specific energy lithium-ion batteries owning to its high theoretical specific capacity, low cost, high operating voltage and environmental friendliness. Yet they suffer from severe capacity and voltage attenuation during prolong cycling, which blocks their commercial application. To clarify these causes, we synthesize $Li_{1.5}Mn_{0.55}Ni_{0.4}Co_{0.05}O_{2.5}$ ($Li_{1.2}Mn_{0.44}Ni_{0.32}Co_{0.04}O_2$) with high-nickel-content cathode material by a solid-sate complexation method, and it manifests a lot slower capacity and voltage attenuation during prolong cycling compared to $Li_{1.5}Mn_{0.66}Ni_{0.17}Co_{0.17}O_{2.5}$ ($Li_{1.2}Mn_{0.54}Ni_{0.13}Co_{0.13}O_2$) and $Li_{1.5}Mn_{0.65}Ni_{0.25}Co_{0.1}O_{2.5}$ ($Li_{1.2}Mn_{0.52}Ni_{0.2}Co_{0.08}O_2$) cathode materials. The capacity retention at 1 C after 100 cycles reaches to 87.5% and the voltage attenuation after 100 cycles is only 0.460 V. Combining X-ray diffraction (XRD), scanning electron microscope (SEM), and transmission electron microscopy (TEM), it indicates that increasing the nickel content not only stabilizes the structure but also alleviates the attenuation of capacity and voltage. Therefore, it provides a new idea for designing of Li-rich layered oxide cathode materials that suppress voltage and capacity attenuation.

Keywords: Li-rich layered oxide; cathode materials; voltage attenuation; lithium-ion batteries; solid-state complexation method

---

1. Introduction

Advanced lithium-ion batteries (LIBs) technology have promoted the rapid development of mobile electronic devices owning to their low cost, long life, lack of any memory effect and environmental friendliness [1–3]. The development of plug in hybrid electric vehicles (PHEVs) and electric vehicles (EVs) puts higher demands on energy density and cruising range [4–6]. Compared to traditional cathode materials, such as $LiCoO_2$ [7,8], spinel $LiMn_2O_4$ [9,10], polyanionic $LiFePO_4$ [11,12], and layered cathode materials $LiMO_2$ [13–16] (M = Ni, Co, Mn), Li-rich layered oxide cathode materials, represented by the chemical formula $xLi_2MnO_3·(1-x)LiMO_2$ (0 < x < 1, M = Ni, Co, Mn) have received extensive attention from scientists all over the world because they exhibit a reversible capacity exceeding 250 mAh·g^{-1} between 2 V and 4.8 V, as well as their low cost and high energy density (>900 Wh·Kg^{-1}) [17,18]. Unfortunately, these cathode materials put up with poor kinetics [19] and severe voltage attenuation [20–22] during prolong cycling, which directly affects their electrochemical performance, particularly the energy that the battery can output [23–25]. These disadvantages hinder the commercial development of high specific energy lithium-ion cells with cathodes prepared with the use of lithium-manganese-rich materials.

In order to explore effective ways to alleviate the voltage attenuation of Li-rich cathode materials, scientists have done a lot of research work to identify the origin [6,26–28]. The voltage attenuation can be caused by voltage fade and increase of resistance during cycling [29]. The cause of voltage attenuation is generally attributed to a continuous phase transition from layered phase to spinel during the repeated extraction/insertion processes [19,21,30], corresponding to the irreversible migration of the transition metal (TM) ions, during the course of which they move from the octahedral sites in the TM slab to the octahedral sites in the Li slab during the lithium ion extraction/insertion processes [31]. It is possible for Li-rich layered oxide cathode materials to prevent collapse of the layered structure by minimizing the tension between the neighboring oxygen layers in its deep delithiation state. In addition, another important reason for the voltage attenuation is increase of resistance during cycling [29]. Therefore, the voltage attenuation caused by resistance growth can be reflected in the average voltage as a function of cycling. Researchers believe that surface structure modification and elemental doping can effectively suppress voltage attenuation. For example, there are many reports of interface modifications (Al_2O_3 [32], $LiFePO_4$ [33], LIPON [32], etc.) will reduce the side reaction between the surface of the positive electrode material and the electrolyte, and enhance the stability of the surface layered structure. elemental doping (Na^+ [34], Ce^{3+}/Ce^{4+}, and Sn^{4+} [22], etc.) may significantly inhibit the Phase transition and stabilize the structure. Furthermore, Burrell et al. explored the effect of cycling temperature on the voltage fade of Li-rich layered oxide cathode materials [35]. Increasing the nickel content can effectively suppress the voltage attenuation in these cobalt-free or high-nickel cathode materials, which is the development trend of lithium-ion batteries in the future [23,25,36]. Dahn et al. reported Core-shell (CS) materials allow that the Mn-rich shell can protect the Ni-rich core from electrolyte attack, as well as the Ni-rich core maintains a high and stable average voltage [27,37]. Li et al. reported that reducing the Co content can significantly suppress voltage attenuation [38]. Vu et al. reported study the effect of composition on the voltage fade of Li-rich cathode materials with a combinatorial synthesis approach. Although the voltage fade can be reduced by controlling the composition of the system, there is no guarantee that the composition is the layered-layered structure [39]. Therefore, it is very important indication for Li-rich layered oxide cathode material to control the composition of TM ions in the $LiMO_2$ layer in order to suppress the voltage attenuation.

To further investigate the effect of different compositions in the $LiMO_2$ layer on the physicochemical properties of Li-rich layered oxide cathode materials, we synthesize $Li_{1.5}Mn_{0.66}Ni_{0.17}Co_{0.17}O_{2.5}$ ($Li_{1.2}Mn_{0.54}Ni_{0.13}Co_{0.13}O_2$) (LL-111), $Li_{1.5}Mn_{0.65}Ni_{0.25}Co_{0.1}O_{2.5}$ ($Li_{1.2}Mn_{0.52}Ni_{0.2}Co_{0.08}O_2$) (LL-523) and $Li_{1.5}Mn_{0.55}Ni_{0.4}Co_{0.05}O_{2.5}$ ($Li_{1.2}Mn_{0.44}Ni_{0.32}Co_{0.04}O_2$) (LL-811) cathode materials by means of solid-state complexation method, and the electrochemical properties have been investigated. During these three composition samples, LL-811 exhibits slower voltage decay and more excellent cycle stability during prolong cycling. The capacity retention rate of the LL-811 is 87.5% at 1 C rate after 100 cycles, and its voltage attenuation is quite low with 0.460 V. The distinctive advantage of this LL-811 Li-rich cathode material may derive from the high nickel content in the layered (R-3m) phase. High-nickel-content Li-rich layered oxide cathode material may cause more Ni^{2+} ions to occupy the Li^+ ion sites in the lithium layer, it can result in a part of nickel to be doped at Li^+ ion sites [40]. The cation doping, to some extent, can improve the structural stability by supporting the Li slabs and reducing tension of neighboring oxygen layers during the delithiation process [23]. In addition, nickel acts as a stabilizer to reduce the complete transformation of manganese by substitution. The preferential reduction of $Ni^{4+/2+}$ maintains average oxidation state of Mn above 3+, as well as suppresses the Jahn-Teller effect caused by Mn^{3+} ions and effectively improves structural durability [41–43].

2. Materials and Methods

2.1. Sample Preparation

The $Li_{1.5}Mn_{0.66}Ni_{0.17}Co_{0.17}O_{2.5}$ ($Li_{1.2}Mn_{0.54}Ni_{0.13}Co_{0.13}O_2$), $Li_{1.5}Mn_{0.65}Ni_{0.25}Co_{0.1}O_{2.5}$ ($Li_{1.2}Mn_{0.52}Ni_{0.2}Co_{0.08}O_2$), $Li_{1.5}Mn_{0.55}Ni_{0.4}Co_{0.05}O_{2.5}$ ($Li_{1.2}Mn_{0.44}Ni_{0.32}Co_{0.04}O_2$) cathode materials (marked as LL-111, LL-523, LL-811, respectively.) were synthesized by a solid-sate complexation method using citric acid monohydrate as complexing agent with analytical grade chemicals $LiAc·2H_2O$ (Excess 5%, AR, 99%), $Ni(Ac)_2·4H_2O$ (AR, 98%), $Co(Ac)_2·4H_2O$ (AR, 99.5%), $Mn(Ac)_2·4H_2O$ (AR, 99%), citric acid monohydrate (AR, 99.5%). The molar ratio between transition metal ion and citric acid monohydrate was 1:1. Using a certain amount of absolute ethanol as a solvent, a stoichiometric amount of above reagents were mixed thoroughly and ball milled at the speed of 200 rpm continuously for 4 h. After ball milling, it was dried in an oven at 80 °C for 24 h in order to get a uniform mixed precursor. The precursor was ball milled for 30 min and precalcined at 450 °C for 6 h in air to eliminate the organic substances then was calcined at 900 °C, 900 °C, 800 °C for 12 h in air at the rate of 5 °C·min^{-1}, respectively. And finally, these required samples were obtained.

2.2. Materials Characterizations

The crystallographic structure LL-111, LL-523 and LL-811 cathode materials were carried out by X-ray diffraction (XRD, Bruker D8, Karlsruhe, Germany) with Cu Kα radiation (λ = 1.54056 Å) in the range of 10–90° with the speed of 5° min^{-1}. The microscopic morphology was investigated with scanning electron microscopy (SEM, TESCAN MIRA3 LMU, Brno, Czech Republic) and transmission electron microscopy (TEM, TECNAI G2 F20, Hillsboro, America).

Electrochemical performance of the samples was characterized using galvanostatic charge-discharge tests with two-electrode coin cells (type CR-2025, Shenzhen, China). All the charge-discharge processes of our CR-2025 cells except for the first cycle was measured under a voltage window of 2–4.6 V. The synthesized sample, acetylene black (AR, Hersbit Chemical Co., Ltd., Shanghai, China) and polyvinylidene difluoride (PVDF, FR905, San ai fu New Material Technology Co., Ltd., Shanghai, China) with a weight ratio of 8:1:1 to make a slurry in the N-methyl pyrrolidone (NMP) solvent. The slurry was uniformly coated onto aluminum foil as current collector and then dried at 120 °C for 6 h under vacuum oven. Cells were assembled in an Argon-filled glove box with H_2O and O_2 contents below 0.01 ppm, using the metallic lithium foil as an anode. The electrolyte was 1 M $LiPF_6$ dissolved in ethyl carbonate (EC) and dimethyl carbonate (DMC) (1:1 in volume) and the separator was Celgard-2500 membrane.

The Galvanostatic charge-discharge measurements were carried out using NEWARE CT-4008 battery testing system (Shenzhen, China) within the voltage range of 2.0–4.8 V at 25 °C. Cyclic voltammetry (CV) and AC impedance (EIS, 1 MHz–0.1 MHz) using a CHI660E electrochemical workstation (Shanghai, China).

3. Results and Discussion

3.1. Crystal Structure and Microstructure

Using citric acid monohydrate as complexing agent, the cathode materials LL-111, LL-523, and LL-811 were synthesized by a solid-state complexation method. The phase and crystal structure of the above cathode materials were respectively analyzed by XRD. The XRD patterns of the sample LL-111, LL-523, LL-811 are shown in Figure 1. Using the JADE 6.5 analysis software to analyze the XRD data, the diffraction peaks of three samples can be indexed to a typical α-$NaFeO_2$ structure (R-3m) and the enlargements of the small region from 20 to 24° are attributed to Li_2MnO_3 phase (C2/m) [44,45]. Complete splitting of the two pair diffraction peaks (006)/(012) and (018)/(110) indicates the integrity of the layered structure for all samples. It is noted that, the intensity of Li_2MnO_3 characteristic peak gradually decreases with the metallic nickel content increasing.

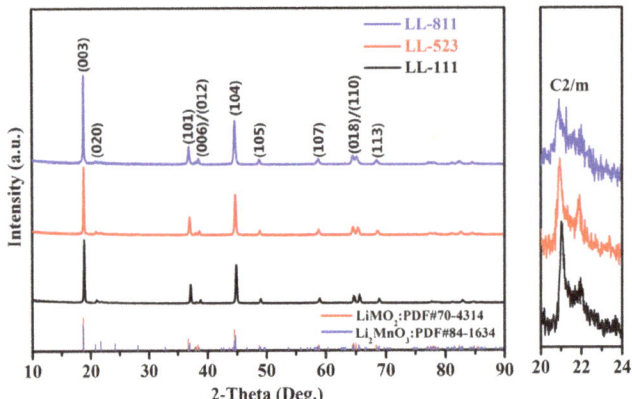

Figure 1. XRD patterns of LL-111, LL-523, LL-811, which adopts a α-NaFeO$_2$ structure. The XRD patterns on the right show the enlargement of Li$_2$MnO$_3$ characteristic peak over a small 2θ region.

The surface morphologies of LL-111, LL-523, LL-811 samples and corresponding element compositions are shown in Figure 2. As seen in Figure 2a–c, three samples show irregular polyhedral morphology with a size of approximately 200–500 nm, which is consistent with previous literatures [6,46,47]. In Figure 2d–f, the primary particles of three samples show different degrees of agglomeration. As the nickel content increases, particle agglomeration becomes more and more seriously. Besides, the interface between the particles is blurred and even disappears. As seen in Figure 2g–i, the actual ratios of LL-111 and LL-523, LL-811 cathode materials are basically in accordance with the designed values. However, LL-811 (Figure 2i) shows a Mn-rich trend on the surface, which may be mainly due to the concentration gradient of Ni/Mn with the nickel content increasing.

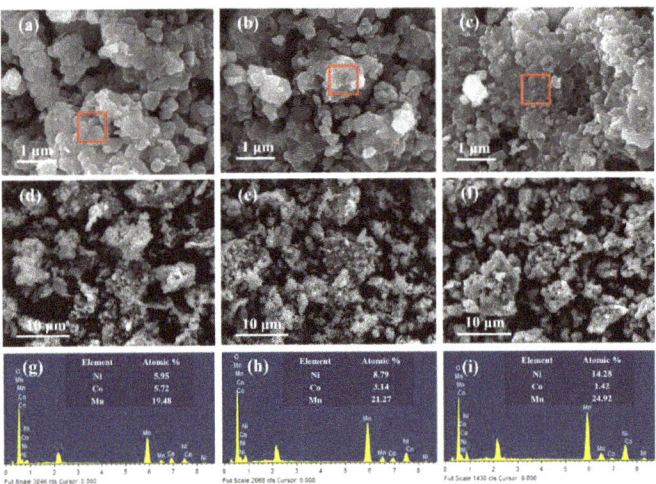

Figure 2. SEM images of LL-111 (**a,d**), LL-523 (**b,e**), LL-811 (**c,f**); Elemental mapping images (**g–i**) of LL-111, LL-523, LL-811, respectively.

To explore the precise structural properties of Li-rich cathode materials with high nickel content, TEM and HRTEM images are shown in Figure 3. As seen in Figure 3a,d,g the edge of LL-811 cathode material is evenly distributed. Besides, it has relatively straight and continuous lattice fringes, and the interplanar spacing of (003) crystal plane is 4.72 Å (Figure 3b), which shows the characteristics of good layered structure material (FFT results in insets of Figure 3c). The lattice fringes of Figure 3e show a distinct two-phase composite (FFT results in insets of Figure 3f). As clearly seen in Figure 3g, the internal phase distribution of the bulk is not very uniform, and Figure 3h further verifies this phenomenon. Figure 3h and FFT results show three kinds of different plane spaces and crystal plane orientations, which represent the (003) crystal plane of the LL-811 structure (4.72 Å), the (006) crystal plane of the LL-811 structure (2.36 Å), two-phase (Li_2MnO_3 and $LiMO_2$) composite, respectively.

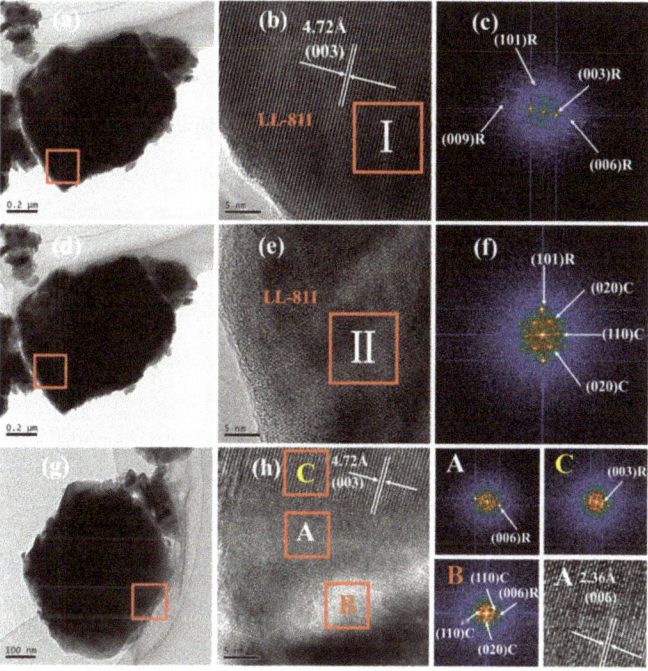

Figure 3. TEM images: (**a,d,g**), HRTEM images: (**b,e,h**) of selected regions of LL-811 and corresponding to the Fast Fourier transform (FFT) images of I and II regions: (**c,f**).

In order to further verify the elemental uniformity of the LL-811 sample, the EDX linear scanning of the single particle was carried out in Figure 4a,b. A very important message can be obtained from Figure 4b, in which the content of Mn element gradually increases from internal to external and Mn-rich phase appears on the surface. However, the change trend of nickel element is opposite, which may be mainly due to the segregation of Mn and Ni elements. The segregation may be induced by citric acid and be also due to the bond formation of the layered material itself [48,49]. Nickel has high catalytic activity and is easy to react with electrolyte, while Mn-rich on the surface can effectively inhibit the reaction. The Mn-rich surface phenomenon may explain the excellent prolong cycling stability of LL-811 cathode materials [25,27].

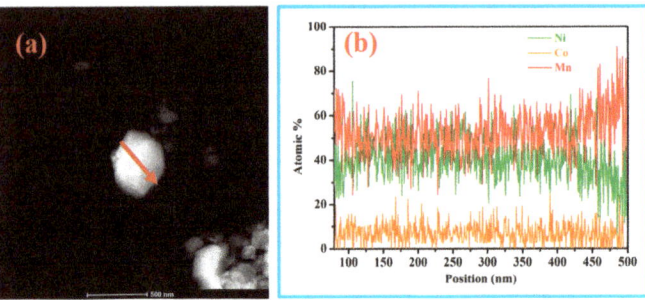

Figure 4. (**a**) TEM image of LL-811 single particle; (**b**) corresponding to the Energy dispersive X-ray linear scanning.

3.2. Electrochemical Charge/Discharge Behavior

In order to evaluate the electrochemical property of the samples LL-111, LL-523, and LL-811, the first charge-discharge curves, rate performances from 0.1 C to 5 C and cycling performances at 1 C (1 C = 200 mAh·g^{-1}) between 2 V to 4.8 V are shown in Figure 5. As shown in Figure 5a, there are two distinct voltage plateaus during the first charge process: (1) a smooth voltage plateau below 4.5 V and (2) a long voltage plateau about 4.5 V [50]. The first discharge capacities of LL-111, LL-523, LL-811 are 284.6 mAh·g^{-1}, 263.0 mAh·g^{-1}, 207.4 mAh·g^{-1}, respectively.

LL-111 samples at 1 C delivers 206.4 mAh·g^{-1} and about 98.5% of the capacity is maintained even after 100 cycles. The first discharge capacity of LL-523 at 1 C is 194.1 mAh·g^{-1} and about 80.0% of the capacity is maintained after 100 cycles. In contrast to LL-111 and LL-523, the capacity retention of the LL-811 is 87.5% after 100 cycles, although the first discharge capacity of LL-811 at 1 C only is 154.6 mAh·g^{-1}. By comparison, the cycle stability of the high-nickel-content LL-811 cathode material is more excellent. It can be attributed to that nickel ions easily migrate out of the TM layer to support the structure instead of being trapped in the middle tetrahedral layer [23].

Figure 5c shows the rate performance of these three samples at different current densities of 0.1 C, 0.2 C, 0.5 C, 1 C, 3 C, 5 C at 25 °C between 2 V and 4.8 V. As seen in Figure 5c, the discharge capacities of LL-111, LL-523, and LL-811 are 130.8 mAh·g^{-1}, 118.9 mAh·g^{-1}, and 100.5 mAh·g^{-1}, respectively at 5 C. The low rate capacity is ascribed to the sluggish kinetics of high nickel Li-rich layered oxide cathode materials discussed in the present paper. The method of calculating the lithium ion diffusion coefficient by EIS measurement is the same as the previous paper [51]. After EIS measurement, Li$^+$ diffusion coefficient of LL-111, LL-523, and LL-811 three samples are 2.77 × 10^{-14} cm^2 S^{-1}, 3.70 × 10^{-14} cm^2 S^{-1}, and 1.32 × 10^{-14} cm^2 S^{-1}, respectively. Moreover, as seen in Figure 5g, the charge transfer impedance (Rct) is 212 Ω, 352 Ω, and 606.7 Ω for the LL-111, LL-523, and LL-811 samples. The Rct of high nickel LL-811 cathode material is very large compared to the other two samples, which can be attributed to the low electronic conductivity.

The most striking performance feature of LL-811 cathode materials is its low voltage attenuation after prolong cycling. Figure 5e shows the relationship between the average voltage and cycle number. The voltage attenuation is approximately 0.460 V after 100 cycles at 1 C, while for LL-111 and LL-523, the voltage attenuation is reached up to 0.665 V and 0.600 V, respectively.

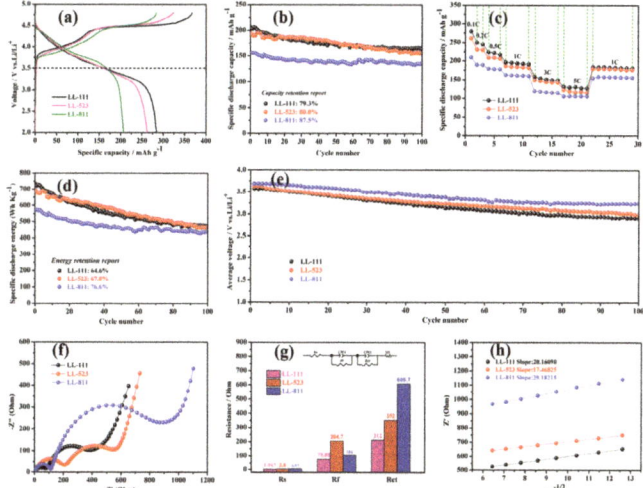

Figure 5. Electrochemical performances: (**a**) First charge- discharge curves at 0.1 C; (**b**) The cycle performances at 1 C; (**c**) The rate performances from 0.1 C to 5 C; (**d**) Specific discharge energy cures at 1 C; (**e**) Average voltage attenuation curves after 100 cycles at 1 C; Nyquist plots: (**f**) Fresh cells; (**g**) Fitted impedance data of LL-111, LL-523, and LL-811 cells between 2 V to 4.8 V; (**h**) The typical plots of Z' vs. $\omega^{-1/2}$ at low frequency.

To further illustrate the voltage attenuation phenomenon of Li-rich cathode materials upon cycling, discharge curves with different cycles between 2 V to 4.8 V at 1 C and cyclic voltammetry (CV) curves of LL-111, LL-523, and LL-811 are shown in Figure 6. As seen in Figure 6a–c, voltage attenuates rapidly at 1 C for LL-111, and for LL-523 it is slower than LL-111, whereas for LL-811 it is the lowest. Figure 6d–f clearly characterize that the three samples display distinct oxidation peaks appeared at 4.0 V during the first cycle, corresponding to the oxidation reactions of $Ni^{2+/3+/4+}$ and $Co^{3+/4+}$; and the oxidation peak at 4.6 V corresponds to the activation process of Li_2MnO_3 [52]. The reduction peaks at 3.3 V, 3.6 V, 4.1 V reflect $Mn^{4+/3+}$, $Ni^{4+/3+}$, $Co^{4+/3+}$ [53], respectively. By comparison, as seen clearly in Figure 6d–f, the voltage decay is minimal for LL-811 after 3 cycles, which is consistent with the phenomenon observed with the voltage attenuation curves. LL-111 and LL-523 Li-rich materials suffers from voltage attenuation after prolong cycling, especially LL-111, for which its capacity mainly came from the low voltage region [23]. Therefore, the specific energy (specific energy = specific capacity × average voltage) output of the battery further lowered upon cycling owing to the disappointing cycle stability and severe voltage attenuation for LL-111 and LL-523 cathode materials. This is due to the phase transformation from layered to spinel-like or rock-salt phases during repeated charge and discharge cycles. As seen in Figure 5d, the absolute value of specific energy of LL-111, LL-523, and LL-811 cathode materials at the 100th cycle is 474.82, 465.88, 435.68 Wh·Kg^{-1}, respectively. Comparing with LL-111 and LL-523 (64.6% and 67.0% energy retention after 100 cycles, respectively), the specific energy retention was 76.7% for LL-811 after 100 cycles. The excellent cell properties of LL-811 indicate that increasing the nickel content could significantly inhibit the intrinsic voltage attenuation of Li-rich materials. These consequences demonstrate that high-nickel-content Li-rich cathode materials, such as LL-811, exhibit outstanding structural durability during prolong cycling, which will promote the commercialization of Li-rich cathode materials.

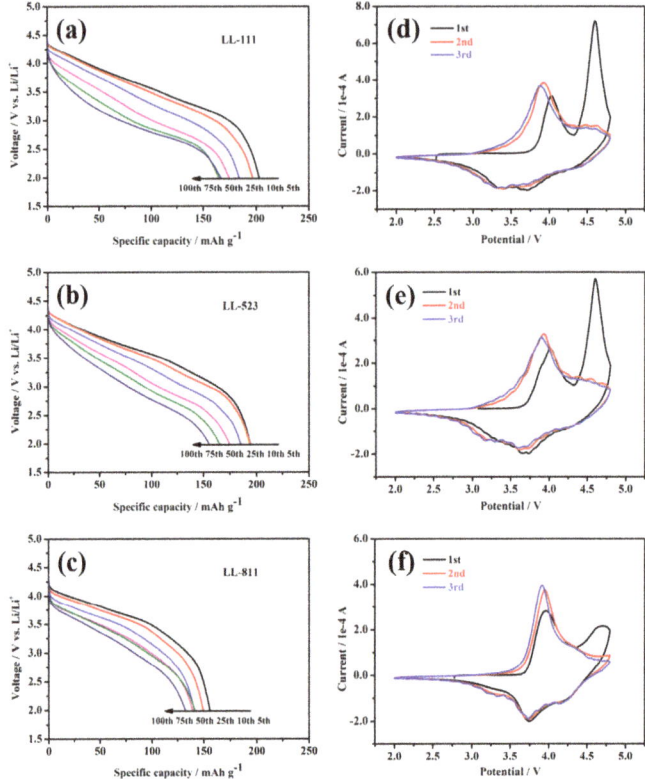

Figure 6. Discharge curves with different cycles between 2 V to 4.8 V at 1 C and Cyclic Voltammetry (CV) curves of LL-111 (**a,d**), LL-523 (**b,e**), and LL-811 (**c,f**).

To further assess the effect of nickel content on the structural durability of Li-rich cathode materials, the XRD patterns of LL-111, LL-523, LL-811 before and after 100 cycles are shown in Figure 7. The intensity ratios of both the (003) and (104) peaks of LL-111, LL-523, and LL-811 are shown in Table 1. Compared to LL-111 and LL-523, the intensity ratio of the (003) and (104) peaks of the LL-811 cathode material is still greater than 1.2 after 100 cycles [54,55], which indicates that the high nickel cathode materials can maintain their structural durability and inhibit the phase transformation to a certain degree during prolong cycling.

In this paper, Li-rich layered oxide cathode materials with different compositions are designed and prepared. From the above discussion results, the sensitivity of the voltage attenuation phenomenon to the composition is relatively large.

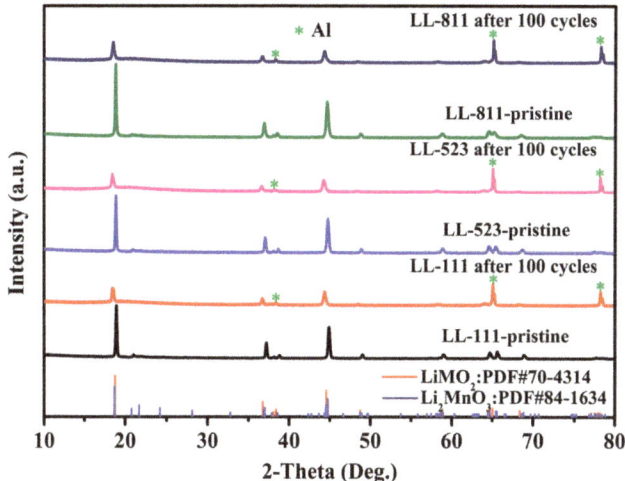

Figure 7. XRD patterns of LL-111, LL-523, LL-811 before and after 100 cycles.

Table 1. The intensity of (003) and (104) and the ratios $I_{(003)}/I_{(104)}$ of LL-111, LL-523, and LL-811 initial samples and cycled electrodes by calculating from the XRD data.

Samples	$I_{(003)}$	$I_{(104)}$	Initial $I_{(003)}/I_{(104)}$	$I_{(003)}$	$I_{(104)}$	Cycled $I_{(003)}/I_{(104)}$
LL-111	6553	3941	1.6628	2215	1803	1.2285
LL-523	7017	4189	1.6751	2286	1611	1.4190
LL-811	9116	4531	2.0120	2670	1563	1.7083

4. Conclusions

In this paper, LL-111, LL-523, and LL-811 cathode materials were successfully synthesized by a solid-sate complexation method using citric acid monohydrate as complexing agent. Compared to the LL-111 and LL-523, the high-nickel-content LL-811 cathode material shows an excellent cycling stability (capacity retention of 87.5% at 1 C rate after 100 cycles and energy retention of 76.7% at 1 C rate after 100 cycles) and suppresses voltage attenuation (only 0.460 V after 100 cycles) during prolong cycling. CV curves also show that the high-nickel-content LL-811 cathode material exhibits less polarization during cycling. What is more, cycled XRD results demonstrate that increasing the nickel content can effectively maintain structural stability. The results show that the sensitivity of the voltage attenuation to the composition is relatively large. This proves that it is very important for Li-rich layered oxide cathode materials to control the composition of TM ions in the $LiMO_2$ layer. This further demonstrates that the nickel content plays a very important role in stabilizing the structure and suppressing the voltage attenuation. This paper provides a reference for the composition design of Li-rich layered oxide materials with the high nickel content.

Author Contributions: Methodology and Conceptualization, Z.C., J.L., and Q.L.; Resources, Z.C. and H.Z.; Data Curation, Y.J., B.L., J.L., and F.L.; Writing-Original Draft Preparation, J.L.; Writing-Review and Editing, Q.L., L.L., and J.D.; Funding Acquisition, Z.C. and H.Z. All authors have read and agreed to the published version of the manuscript.

Funding: This work was supported by the National Natural Science Foundation of China (No. 51874048), the National Science Foundation for Young Scientists of China (No. 51604042), the Research Foundation of Education Bureau of Hunan Province (No. 19A003) and Scientific Research Fund of Changsha Science and Technology Bureau (No. kq1901100) and Postgraduate Innovative Test Program of Hunan Province.

Conflicts of Interest: The authors declare no conflict of interest.

References

1. Li, J.F.; Li, M.; Zhang, L.; Wang, J.Z. General synthesis of $x\text{Li}_2\text{MnO}_3\cdot(1-x)\text{LiNi}_{1/3}\text{Co}_{1/3}\text{Mn}_{1/3}\text{O}_2$ (x = 1/4, 1/3, and 1/2) hollow microspheres towards enhancing the performance of rechargeable lithium ion batteries. *J. Mater. Chem. A* **2016**, *4*, 12442–12450. [CrossRef]
2. Chen, Z.Y.; Yan, X.Y.; Xu, M.; Cao, K.F.; Zhu, H.L.; Li, L.J.; Duan, J.F. Building Honeycomb-Like Hollow Microsphere Architecture in a Bubble Template Reaction for High-Performance Lithium-Rich Layered Oxide Cathode Materials. *ACS Appl. Mater. Interfaces* **2017**, *9*, 30617–30625. [CrossRef]
3. Yang, H.P.; Wu, H.H.; Ge, M.Y.; Li, L.J.; Yuan, Y.F.; Yao, Q.; Chen, J.; Xia, L.F.; Zheng, J.M.; Chen, Z.Y.; et al. Simultaneously Dual Modification of Ni-Rich Layered Oxide Cathode for High-Energy Lithium-Ion Batteries. *Adv. Funct. Mater.* **2019**, *29*, 1808825–1808837. [CrossRef]
4. Whittingham, M.S. Lithium Batteries and Cathode Materials. *Chem. Rev.* **2004**, *104*, 4271–4302. [CrossRef]
5. Armand, M.; Tarascon, J.M. Building better batteries. *Nature* **2008**, *451*, 652–657. [CrossRef] [PubMed]
6. Xu, M.; Chen, Z.Y.; Li, L.J.; Zhu, H.L.; Zhao, Q.F.; Xu, L.; Peng, N.F.; Gong, L. Highly crystalline alumina surface coating from hydrolysis of aluminum isopropoxide on lithium-rich layered oxide. *J. Power Sources* **2015**, *281*, 444–454. [CrossRef]
7. Hu, B.; Lou, X.B.; Li, C.; Geng, F.S.; Zhao, C.; Wang, J.Y.; Shen, M.; Hu, B.W. Reversible phase transition enabled by binary Ba and Ti-based surface modification for high voltage LiCoO_2 cathode. *J. Power Sources* **2019**, *438*, 226954–226961. [CrossRef]
8. Choi, J.W.; Aurbach, D. Promise and reality of post-lithium-ion batteries with high energy densities. *Nat. Rev. Mater.* **2016**, *1*, 1–16. [CrossRef]
9. Huang, W.; Wang, G.; Luo, C.; Xu, Y.B.; Xu, Y.; Ecksteina, B.J.; Chen, Y.; Wang, B.H.; Huang, J.X.; Kang, Y.J.; et al. Controllable growth of LiMn_2O_4 by carbohydrate-assisted combustion synthesis for high performance Li-ion batteries. *Nano Energy* **2019**, *64*, 103936–103945. [CrossRef]
10. Wang, Y.H.; Wang, Y.H.; Jia, D.S.; Peng, Z.; Xia, Y.Y.; Zheng, G.F. All-nanowire based Li-ion full cells using homologous Mn_2O_3 and LiMn_2O_4. *Nano Lett.* **2014**, *14*, 1080–1084. [CrossRef]
11. Gu, L.; Zhu, C.B.; Li, H.; Yu, Y.; Li, C.L.; Tsukimoto, S.; Maier, J.; Ikuhara, Y. Direct observation of lithium staging in partially delithiated LiFePO_4 at atomic resolution. *J. Am. Chem. Soc.* **2011**, *133*, 4661–4663. [CrossRef] [PubMed]
12. Zhao, Y.; Peng, L.L.; Liu, B.R.; Yu, G.H. Single-crystalline LiFePO_4 nanosheets for high-rate Li-ion batteries. *Nano Lett.* **2014**, *14*, 2849–2853. [CrossRef] [PubMed]
13. Myung, S.T.; Maglia, F.; Park, K.J.; Yoon, C.S.; Lamp, P.; Kim, S.J.; Sun, Y.K. Nickel-Rich Layered Cathode Materials for Automotive Lithium-Ion Batteries: Achievements and Perspectives. *ACS Energy Lett.* **2016**, *2*, 196–223. [CrossRef]
14. Yang, H.; Du, K.; Hu, G.R.; Peng, Z.D.; Cao, Y.B.; Wu, K.P.; Lu, Y.; Qi, X.Y.; Mu, K.C.; Wu, J.L. Graphene@TiO_2 co-modified $\text{LiNi}_{0.6}\text{Co}_{0.2}\text{Mn}_{0.2}\text{O}_2$ cathode materials with enhanced electrochemical performance under harsh conditions. *Electrochim. Acta* **2018**, *289*, 149–157. [CrossRef]
15. Chen, Z.Y.; Gong, X.L.; Zhu, H.L.; Cao, K.F.; Liu, Q.M.; Liu, J.; Li, L.J.; Duan, J.F. High Performance and Structural Stability of K and Cl Co-Doped $\text{LiNi}_{0.5}\text{Co}_{0.2}\text{Mn}_{0.3}\text{O}_2$ Cathode Materials in 4.6 Voltage. *Front. Chem.* **2018**, *6*, 643–668. [CrossRef]
16. Chen, Z.Y.; Cao, K.F.; Zhu, H.L.; Li, L.J.; Gong, X.L.; Liu, Q.M.; Duan, J.F. Improved Electrochemical Performance of Surface Coated $\text{LiNi}_{0.80}\text{Co}_{0.15}\text{Al}_{0.05}\text{O}_2$ With Polypyrrole. *Front. Chem.* **2018**, *6*, 648–657. [CrossRef] [PubMed]
17. Yabuuchi, N.; Yoshii, K.; Myung, S.T.; Nakai, I.; Komaba, S. Detailed studies of a high-capacity electrode material for rechargeable batteries, Li_2MnO_3-$\text{LiCo}_{1/3}\text{Ni}_{1/3}\text{Mn}_{1/3}\text{O}_2$. *J. Am. Chem. Soc.* **2011**, *133*, 4404–4419. [CrossRef]
18. Ma, Y.T.; Liu, P.F.; Xie, Q.S.; Zhang, G.B.; Zheng, H.F.; Cai, Y.X.; Li, Z.; Wang, L.S.; Zhu, Z.Z.; Mai, L.Q.; et al. Double-shell Li-rich layered oxide hollow microspheres with sandwich-like carbon@spinel@layered@spinel@carbon shells as high-rate lithium ion battery cathode. *Nano Energy* **2019**, *59*, 184–196. [CrossRef]
19. Qing, R.P.; Shi, J.L.; Xiao, D.D.; Zhang, X.D.; Yin, Y.X.; Zhai, Y.B.; Gu, L.; Guo, Y.G. Enhancing the Kinetics of Li-Rich Cathode Materials through the Pinning Effects of Gradient Surface Na^+ Doping. *Adv. Energy Mater.* **2016**, *6*, 1501914–1501919. [CrossRef]
20. Ning, F.H.; Shang, H.F.; Li, B.; Jiang, N.; Zou, R.Q.; Xia, D.G. Surface thermodynamic stability of Li-rich Li_2MnO_3: Effect of defective graphene. *Energy Storage Mater.* **2019**, *22*, 113–119. [CrossRef]

21. Yan, P.F.; Nie, A.M.; Zheng, J.M.; Zhou, Y.G.; Lu, D.P.; Zhang, X.F.; Xu, R.; Belharouak, I.; Zu, X.T.; Xiao, J.; et al. Evolution of lattice structure and chemical composition of the surface reconstruction layer in $Li_{1.2}Ni_{0.2}Mn_{0.6}O_2$ cathode material for lithium ion batteries. *Nano Lett.* **2015**, *15*, 514–522. [CrossRef] [PubMed]
22. Liu, Y.Y.; Li, R.R.; Li, J.L.; Yang, Z.; Zhong, J.J.; Wang, Z.; Kang, F.Y. A high-performance Ce and Sn co-doped cathode material with enhanced cycle performance and suppressed voltage decay for lithium ion batteries. *Ceram. Int.* **2019**, *45*, 20780–20787. [CrossRef]
23. Shi, J.L.; Zhang, J.N.; He, M.; Zhang, X.D.; Yin, Y.X.; Li, H.; Guo, Y.G.; Gu, L.; Wan, L.J. Mitigating Voltage Decay of Li-Rich Cathode Material via Increasing Ni Content for Lithium-Ion Batteries. *ACS Appl. Mater. Interfaces* **2016**, *8*, 20138–20146. [CrossRef] [PubMed]
24. Xu, M.; Chen, Z.Y.; Zhu, H.L.; Yan, X.Y.; Li, L.J.; Zhao, Q.F. Mitigating capacity fade by constructing highly ordered mesoporous Al_2O_3/polyacene double-shelled architecture in Li-rich cathode materials. *J. Mater. Chem. A* **2015**, *3*, 13933–13945. [CrossRef]
25. Ju, X.K.; Hou, X.; Liu, Z.Q.; Zheng, H.F.; Huang, H.; Qu, B.H.; Wang, T.H.; Li, Q.H.; Li, J. The full gradient design in Li-rich cathode for high performance lithium ion batteries with reduced voltage decay. *J. Power Sources* **2019**, *437*, 226902–226910. [CrossRef]
26. Gu, M.; Genc, A.; Belharouak, I.; Wang, D.P.; Amine, K.; Thevuthasan, S.; Baer, D.R.; Zhang, J.G.; Browning, N.D.; Liu, J.; et al. Nanoscale Phase Separation, Cation Ordering, and Surface Chemistry in Pristine $Li_{1.2}Ni_{0.2}Mn_{0.6}O_2$ for Li-Ion Batteries. *Chem. Mater.* **2013**, *25*, 2319–2326. [CrossRef]
27. Li, J.; Camardese, J.; Shunmugasundaram, R.; Glazier, S.; Lu, Z.G.; Dahn, J.R. Synthesis and Characterization of the Lithium-Rich Core–Shell Cathodes with Low Irreversible Capacity and Mitigated Voltage Fade. *Chem. Mater.* **2015**, *27*, 3366–3377. [CrossRef]
28. Sathiya, M.; Abakumov, A.M.; Foix, D.; Rousse, G.; Ramesha, K.; Saubanère, M.; Doublet, M.L.; Vezin, H.; Laisa, C.P.; Prakash, A.S.; et al. Origin of voltage decay in high-capacity layered oxide electrodes. *Nat. Mater.* **2014**, *14*, 230–238. [CrossRef]
29. Bettge, M.; Li, Y.; Gallagher, K.; Zhu, Y.; Wu, Q.L.; Lu, W.Q.; Bloom, I.; Abrahama, P.D. Voltage Fade of Layered Oxides Its Measurement and Impact on Energy Density. *J. Electrochem. Soc.* **2013**, *160*, A2046–A2055. [CrossRef]
30. Boulineau, A.; Simonin, L.; Colin, J.F.; Canévet, E.; Daniel, L.; Patoux, S. Evolutions of $Li_{1.2}Mn_{0.61}Ni_{0.18}Mg_{0.01}O_2$ during the Initial Charge/Discharge Cycle Studied by Advanced Electron Microscopy. *Chem. Mater.* **2012**, *24*, 3558–3566. [CrossRef]
31. Wu, F.; Li, N.; Su, Y.F.; Zhang, L.J.; Bao, L.Y.; Wang, J.; Chen, L.; Zheng, Y.; Dai, L.Q.; Peng, J.Y.; et al. Ultrathin spinel membrane-encapsulated layered lithium-rich cathode material for advanced Li-ion batteries. *Nano Lett.* **2014**, *14*, 3550–3555. [CrossRef] [PubMed]
32. Bloom, I.; Trahey, L.; Abouimrane, A.; Belharouak, I.; Zhang, X.F.; Wu, Q.L.; Lu, W.Q.; Abraham, D.P.; Bettge, M.; Elam, J.W.; et al. Effect of interface modifications on voltage fade in $0.5Li_2MnO_3 \cdot 0.5LiNi_{0.375}Mn_{0.375}Co_{0.25}O_2$ cathode materials. *J. Power Sources* **2014**, *249*, 509–514. [CrossRef]
33. Zheng, F.; Yang, C.; Xiong, X.; Xiong, J.; Hu, R.; Chen, Y.; Liu, M. Nanoscale Surface Modification of Lithium-Rich Layered-Oxide Composite Cathodes for Suppressing Voltage Fade. *Angew. Chem. Int. Ed. Engl.* **2015**, *54*, 13058–13062. [CrossRef]
34. Tian, Z.X.; Wang, J.L.; Liu, S.Z.; Li, Q.; Zeng, G.F.; Yang, Y.; Cui, Y.H. Na-stabilized Ru-based lithium rich layered oxides with enhanced electrochemical performance for lithium ion batteries. *Electrochim. Acta* **2017**, *253*, 31–38. [CrossRef]
35. Vu, A.; Walker, L.K.; Bareno, J.; Burrell, A.K.; Bloom, I. Effects of cycling temperatures on the voltage fade phenomenon in $0.5Li_2MnO_3 \cdot 0.5LiNi_{0.375}Mn_{0.375}CO_{0.25}O_2$ cathodes. *J. Power Sources* **2015**, *280*, 155–158. [CrossRef]
36. Li, H.Y.; Cormier, M.; Zhang, N.; Inglis, J.; Li, J.; Dahn, J.R. Is Cobalt Needed in Ni-Rich Positive Electrode Materials for Lithium Ion Batteries? *J. Electrochem. Soc.* **2019**, *166*, A429–A439. [CrossRef]
37. Kim, U.H.; Kim, J.H.; Hwang, J.Y.; Ryu, H.H.; Yoon, C.S.; Sun, Y.K. Compositionally and structurally redesigned high-energy Ni-rich layered cathode for next-generation lithium batteries. *Mater. Today* **2019**, *23*, 26–36. [CrossRef]
38. Wang, J.; He, X.; Paillard, E.; Laszczynski, N.; Li, J.; Passerini, S. Lithium- and Manganese-Rich Oxide Cathode Materials for High-Energy Lithium Ion Batteries. *Adv. Energy Mater.* **2016**, *6*, 1600906–1600922. [CrossRef]

39. Vu, A.; Qin, Y.; Lin, C.K.; Abouimrane, A.; Burrell, A.K.; Bloom, S.; Bass, D.; Bareno, J.; Bloom, I. Effect of composition on the voltage fade phenomenon in lithium-, manganese-rich xLiMnO$_3$·(1-x)LiNi$_a$Mn$_b$Co$_c$O$_2$: A combinatorial synthesis approach. *J. Power Sources* **2015**, *294*, 711–718. [CrossRef]
40. Shi, J.L.; Xiao, D.D.; Ge, M.; Yu, X.; Chu, Y.; Huang, X.; Zhang, X.D.; Yin, Y.X.; Yang, X.Q.; Guo, Y.G.; et al. High-Capacity Cathode Material with High Voltage for Li-Ion Batteries. *Adv. Mater.* **2018**, *30*, 1705575–1705582. [CrossRef]
41. Hy, S.; Cheng, J.H.; Liu, J.Y.; Pan, C.J.; Rick, J.; Lee, J.F.; Chen, J.M.; Hwang, B.J. Understanding the Role of Ni in Stabilizing the Lithium-Rich High-Capacity Cathode Material Li[Ni$_x$Li$_{(1-2x)/3}$Mn$_{(2-x)/3}$]O$_2$(0 ≤ x ≤ 0.5). *Chem. Mater.* **2014**, *26*, 6919–6927. [CrossRef]
42. Kim, T.; Song, B.H.; Lunt, A.G.; Cibin, G.; Dent, A.J.; Lu, L.; Korsunsky, A.M. In operando X-ray absorption spectroscopy study of charge rate effects on the atomic environment in graphene-coated Li-rich mixed oxide cathode. *Mater. Des.* **2016**, *98*, 231–242. [CrossRef]
43. Yang, F.; Zhang, Q.G.; Hu, X.H.; Peng, T.Y.; Liu, J.Q. Preparation of Li-rich layered-layered type xLi$_2$MnO$_3$·(1− x) LiMnO$_2$ nanorods and its electrochemical performance as cathode material for Li-ion battery. *J. Power Sources* **2017**, *353*, 323–332. [CrossRef]
44. Yan, P.F.; Xiao, L.; Zheng, J.M.; Zhou, Y.G.; He, Y.; Zu, X.T.; Mao, S.X.; Xiao, J.; Gao, F.; Zhang, J.G.; et al. Probing the Degradation Mechanism of Li$_2$MnO$_3$ Cathode for Li-Ion Batteries. *Chem. Mater.* **2015**, *27*, 975–982. [CrossRef]
45. Wang, P.B.; Luo, M.Z.; Zheng, J.C.; He, Z.J.; Tong, H.; Yu, W.J. Comparative Investigation of 0.5Li$_2$MnO$_3$·0.5LiNi$_{0.5}$Co$_{0.2}$Mn$_{0.3}$O$_2$ Cathode Materials Synthesized by Using Different Lithium Sources. *Front. Chem.* **2018**, *6*, 159–167. [CrossRef]
46. Zhang, L.J.; Wu, B.R.; Li, N.; Wu, F. Hierarchically porous micro-rod lithium-rich cathode material Li$_{1.2}$Ni$_{0.13}$Mn$_{0.54}$Co$_{0.13}$O$_2$ for high performance lithium-ion batteries. *Electrochim. Acta* **2014**, *118*, 67–74. [CrossRef]
47. Zhou, L.Z.; Xu, Q.J.; Liu, M.S.; Jin, X. Novel solid-state preparation and electrochemical properties of Li$_{1.13}$[Ni$_{0.2}$Co$_{0.2}$Mn$_{0.47}$]O$_2$ material with a high capacity by acetate precursor for Li-ion batteries. *Solid State Ion.* **2013**, *249–250*, 134–138. [CrossRef]
48. Kandhasamy, S.; Pandey, A.; Minakshi, M. Polyvinylpyrrolidone assisted sol-gel route LiCo$_{1/3}$Mn$_{1/3}$Ni$_{1/3}$PO$_4$ composite cathode for aqueous rechargeable battery. *Electrochim. Acta* **2012**, *60*, 170–176. [CrossRef]
49. Navaratnarajah, K.; Efstratia, S.; Yerassimos, P.; Alexander, C. Defect Process, Dopant Behaviour and Li Ion Mobility in the Li$_2$MnO$_3$ Cathode Material. *Energies* **2019**, *12*, 1329.
50. Yang, S.Q.; Wang, P.B.; Wei, H.X.; Tang, L.B.; Zhang, X.H.; He, Z.J.; Li, Y.J.; Tong, H.; Zheng, J.C. Li$_4$V$_2$Mn(PO$_4$)$_4$-stablized Li[Li$_{0.2}$Mn$_{0.54}$Ni$_{0.13}$Co$_{0.13}$]O$_2$ cathode materials for lithium ion batteries. *Nano Energy* **2019**, *63*, 103889–103898. [CrossRef]
51. Redel, K.; Kulka, A.; Plewa, A.; Molenda, J. High-Performance Li-Rich Layered Transition Metal Oxide Cathode Materials for Li-Ion Batteries. *J. Electrochem. Soc.* **2019**, *166*, A5333–A5342. [CrossRef]
52. Zuo, Y.X.; Li, B.; Jiang, N.; Chu, W.S.; Zhang, H.; Zou, R.Q.; Xia, D.G. A High-Capacity O2-Type Li-Rich Cathode Material with a Single-Layer Li$_2$MnO$_3$ Superstructure. *Adv. Mater.* **2018**, *30*, 1707255–1707259. [CrossRef] [PubMed]
53. Xiong, F.Y.; Tan, S.H.; Wei, Q.L.; Zhang, G.B.; Sheng, J.Z.; An, Q.Y.; Mai, L.Q. Three-dimensional graphene frameworks wrapped Li$_3$V$_2$(PO$_4$)$_3$ with reversible topotactic sodium-ion storage. *Nano Energy* **2017**, *32*, 347–352. [CrossRef]
54. Liu, J.L.; Chen, L.; Hou, M.Y.; Wang, F.; Che, R.C.; Xia, Y.Y. General synthesis of xLi$_2$MnO$_3$·(1-x) LiMn$_{1/3}$Ni$_{1/3}$Co$_{1/3}$O$_2$ nanomaterials by a molten-salt method: Towards a high capacity and high power cathode for rechargeable lithium batteries. *J. Mater. Chem.* **2012**, *22*, 25380–25387. [CrossRef]
55. Yang, X.K.; Wang, D.; Yu, R.Z.; Bai, Y.S.; Shu, H.B.; Ge, L.; Guo, H.P.; Wei, Q.L.; Liu, L.; Wang, X.Y. Suppressed capacity/voltage fading of high-capacity lithium-rich layered materials via the design of heterogeneous distribution in the composition. *J. Mater. Chem. A* **2014**, *2*, 3899–3911. [CrossRef]

© 2019 by the authors. Licensee MDPI, Basel, Switzerland. This article is an open access article distributed under the terms and conditions of the Creative Commons Attribution (CC BY) license (http://creativecommons.org/licenses/by/4.0/).

Article

High-Performance Lithium-Rich Layered Oxide Material: Effects of Preparation Methods on Microstructure and Electrochemical Properties

Qiming Liu [1], Huali Zhu [2], Jun Liu [1], Xiongwei Liao [1], Zhuolin Tang [1], Cankai Zhou [1], Mengming Yuan [1], Junfei Duan [1], Lingjun Li [1] and Zhaoyong Chen [1,*]

[1] College of Materials Science and Engineering, Changsha University of Science and Technology, Changsha 410114, China; liuqiming@stu.csust.edu.cn (Q.L.); liujun@stu.csust.edu.cn (J.L.); avanda666@163.com (X.L.); tzl_edu29@163.com (Z.T.); zhoucankai@stu.csust.edu.cn (C.Z.); yuanmengming@stu.csust.edu.cn (M.Y.); junfei_duan@csust.edu.cn (J.D.); lingjun.li@csust.edu.cn (L.L.)

[2] School of Physics and Electronic Science, Changsha University of Science and Technology, Changsha 410114, China; juliezhu2005@126.com

* Correspondence: chenzhaoyongcioc@126.com; Tel.: +86-137-8711-2902

Received: 10 December 2019; Accepted: 9 January 2020; Published: 11 January 2020

Abstract: Lithium-rich layered oxide is one of the most promising candidates for the next-generation cathode materials of high-energy-density lithium ion batteries because of its high discharge capacity. However, it has the disadvantages of uneven composition, voltage decay, and poor rate capacity, which are closely related to the preparation method. Here, $0.5Li_2MnO_3 \cdot 0.5LiMn_{0.8}Ni_{0.1}Co_{0.1}O_2$ was successfully prepared by sol–gel and oxalate co-precipitation methods. A systematic analysis of the materials shows that the $0.5Li_2MnO_3 \cdot 0.5LiMn_{0.8}Ni_{0.1}Co_{0.1}O_2$ prepared by the oxalic acid co-precipitation method had the most stable layered structure and the best electrochemical performance. The initial discharge specific capacity was 261.6 mAh·g^{-1} at 0.05 C, and the discharge specific capacity was 138 mAh·g^{-1} at 5 C. The voltage decay was only 210 mV, and the capacity retention was 94.2% after 100 cycles at 1 C. The suppression of voltage decay can be attributed to the high nickel content and uniform element distribution. In addition, tightly packed porous spheres help to reduce lithium ion diffusion energy and improve the stability of the layered structure, thereby improving cycle stability and rate capacity. This conclusion provides a reference for designing high-energy-density lithium-ion batteries.

Keywords: lithium-rich layered oxide; cathode material; $0.5Li_2MnO_3 \cdot 0.5LiMn_{0.8}Ni_{0.1}Co_{0.1}O_2$; voltage decay; co-precipitation method; sol–gel method

---

1. Introduction

With the rapid development of hybrid vehicles and electric vehicles, lithium-ion batteries have been more widely used [1–4]. However, the traditional lithium-ion battery cathode materials (e.g., $LiMn_2O_4$, $LiFePO_4$, $LiNi_{1/3}Co_{1/3}Mn_{1/3}O_2$, $LiNi_{0.5}Co_{0.2}Mn_{0.3}O_2$) have a discharge capacity of less than 200 mAh·g^{-1}, which cannot meet the development requirements of the electric vehicle industry [5–7]. The current solution is to increase the nickel content or use lithium-rich layered oxides to increase the discharge capacity [8–12]. The lithium-rich layered oxide has become a candidate for the next generation of high-energy density lithium-ion battery cathode materials due to its high capacity and high operating voltage [13,14]. However, before commercialization, the technical challenge of voltage decay must be addressed [15–18].

Many studies have shown that ion doping [19–21] and surface coating [22–24] can suppress voltage decay of lithium-rich layered oxides. In addition, increasing the nickel content in the lithium-rich layered oxide can also significantly suppress the voltage decay, and the energy density is also improved [25–27]. The preparation methods of lithium-rich layered oxides include solid-state [28], co-precipitation [25], sol–gel [29], solvothermal [30], freeze drying [31], and bubble template [32]. It is controversial whether the lithium-rich layered oxide is a solid solution or a two-phase composite structure, and the structure and performance of the lithium-rich layered oxide are closely related to the synthesis [33–35]. In addition, the preparation method also affects the atomic spatial uniformity of chemical species [36]. The effect of lithium-rich layered oxides with low nickel content on the synthesis method has been studied intensively [37–39]. However, for lithium-rich layered oxides with high nickel content, such as $0.5Li_2MnO_3 \cdot 0.5LiNi_{0.8}Co_{0.1}Mn_{0.1}O_2$ (LL-811), the effects of the preparation method on the microstructure and electrochemical performance and the reasons for the decreased discharge capacity with increasing nickel content have not been investigated.

In this work, we compared the effects of the sol–gel method and the oxalate co-precipitation method on the microstructure, element distribution, and electrochemical performance of LL-811. Two kinds of typical chelating agents, citric acid and sucrose, were selected in the sol–gel method. LL-811 prepared by the oxalate co-precipitation method had the best comprehensive performance. After 100 cycles at 1 C, the voltage decay was 210 mV, and the capacity retention was 94.2%. The discharge specific capacity still reached 138 mAh·g^{-1} at 5 C. The significant reduction in voltage decay can be attributed to the high nickel content and uniform element distribution. In addition, tightly packed porous spheres contributed to reducing the lithium ion diffusion energy barrier and improving cycle stability and rate capacity.

2. Experiment

2.1. The Reagents and Materials

Lithium acetate dihydrate (Li(CH$_3$COO)·2H$_2$O, 99.0%), manganese acetate tetrahydrate (Mn(CH$_3$COO)$_2$·4H$_2$O, 99.0%), cobalt acetate tetrahydrate (Co(CH$_3$COO)$_2$·4H$_2$O, 99.5%), nickel acetate tetrahydrate (Ni(CH$_3$COO)$_2$·4H$_2$O, 98.0%), ammonium oxalate monohydrate (C$_2$H$_8$N$_2$O$_4$·H$_2$O, 99.5%), oxalic acid dihydrate (C$_2$H$_2$O$_4$·2H$_2$O, 99.5%), sucrose (C$_{12}$H$_{22}$O$_{11}$, 99.0%), citric acid monohydrate (C$_6$H$_8$O$_7$·H$_2$O, 99.5%), sodium hydroxide (NaOH, 96.0%), nitric acid (HNO$_3$, 65–68%), and ammonium hydroxide aqueous solution (NH$_3$·H$_2$O, 26.7%) were purchased from Sinopharm Chemical Reagent (Shanghai, China). Lithium hydroxide monohydrate (LiOH·H$_2$O, 98.0%) was purchased from Xilong Scientific Co., Ltd. (Shantou, China). Nickel sulfate hexahydrate (NiSO$_4$·6H$_2$O, 22 wt%) was purchased from Jinchuan group Co., Ltd. (Jinchang, China). Cobalt sulfate heptahydrate (CoSO$_4$·4H$_2$O, 21 wt%) was purchased from Huayou Cobalt Co., Ltd. (Jiangxing, China) Manganese sulfate monohydrate (MnSO$_4$·H$_2$O, 31.8 wt%) was purchased from ISKY Chemicals Co., Ltd. (Changsha, China).

2.2. Sol–Gel Method

LL-811 was synthesized by sol–gel method, using acetic acid salts as raw materials and citric acid monohydrate as a chelating agent. Firstly, according to the stoichiometric ratio, lithium acetate dihydrate, nickel acetate tetrahydrate, cobalt acetate tetrahydrate, and manganese acetate tetrahydrate were added to deionized water to prepare a solution A of 1.5 mol/L. citric acid monohydrate was added to deionized water and configured to a concentration of 2 mol/L solution B. The molar ratio of metal ion to chelating agent was fixed to be 1:1.5. Then, the A and B solutions were slowly mixed, and the pH of the solution was adjusted to 7–8 with ammonium hydroxide. The resulting solution was evaporated at 80 °C in a constant temperature water bath to form a transparent xerogel. The gel was dried in a forced air oven at 120 °C for 12 h to obtain a precursor. Finally, the precursor was calcined at 480 °C for 5 h and then calcined at 850 °C for 12 h in air to obtain a lithium-rich layered oxide (labeled as SLC). When using the same procedure, the chelating agent was changed to sucrose, and the pH was adjusted to 5 with dilute nitric acid. The acquired sample was labeled as SLS.

2.3. Co-Precipitation Method

LL-811 was synthesized by oxalate co-precipitation. Firstly, according to the stoichiometric ratio, nickel sulfate hexahydrate, cobalt sulfate heptahydrate, and manganese sulfate monohydrate were added to deionized water to prepare a solution C of 1.0 mol/L. Ammonium oxalate monohydrate and oxalic acid dihydrate were added to deionized water to configure a solution D with an oxalate concentration of 0.6 mol/L. The ratios of the amounts of the transition metal salts to the substances of the complexing agent and the precipitating agent were 1:2 and 1:1.2, respectively. Secondly, the solution C and the solution D were simultaneously slowly pumped into a 5 L continuous stirred-tank reactor (CSTR). The temperature and agitation speed were maintained at 50 °C and 800 rpm, respectively. In the co-precipitation process, the pH of the solution was maintained at 6.6 by the addition of 1 mol/L sodium hydroxide solution. After the feed was completed, stirring was continued for 1 h to allow the metal ions to completely precipitate. In order to prevent oxidation of the metal cations throughout the process, the CSTR was maintained under a nitrogen atmosphere. Finally, the prepared oxalate precursor was acquired by vacuum filtration and washed with deionized water and then dried in a drying oven at 80 °C for 12 h. The prepared oxalate precursor was thoroughly mixed with 5 wt% excess of $LiOH \cdot H_2O$ powder. The mixture was transferred to a furnace and calcined at 480 °C for 5 h and then calcined at 850 °C for 12 h in air to obtain a lithium-rich layered oxide (labeled as OCP).

2.4. Materials Characterization

The crystal structure of LL-811 was observed with an X-ray diffractometer (XRD, Bruker AXS D8 Advance, Bruker Corporation, Karlsruhe, Germany). The morphology was characterized by field emission scanning electron microscopy (SEM, TESCAN MIRA3 LMU, TESCAN, Brno, Czech Republic). The elemental compositions were characterized using energy dispersive spectroscopy (EDS, Oxford X-Max20, Oxford, UK). The analysis of the microstructure and composition of LL-811 was performed on a transmission electron microscope (TEM, Tecnai G2 F20, Hillsboro, OR, USA) equipped with energy dispersive X-rays (EDX).

The preparation of the positive electrode sheet, the assembly of the button cell, and the test procedures for electrochemical performance were consistent with the previously published articles [40]. The electrolyte is 1 M $LiPF_6$, which is soluble in ethyl carbonate (EC) and dimethyl carbonate (DMC) (1:1 by volume, CAPCHEM, Shenzhen, China). The load of the active material on the positive electrode sheet was about 2.7 mg·cm^{-2}. The CT-4008 battery testing system of NEWARE (Shenzhen, China) was used to test charge and discharge performance of the battery. Electrochemical impedance spectroscopy was performed with a CHI660E electrochemical workstation (Chinstruments, Shanghai, China) in the frequency range of 1 mHz to 100 kHz.

3. Results and Discussion

The powder X-ray diffraction patterns of the lithium-rich layered oxide prepared by two different methods are shown in Figure 1. The major peaks in the XRD pattern could be indexed based on α-NaFeO$_2$ layered structure with space group R$\bar{3}$m and monoclinic symmetry with space group C2/m [32]. The (006)/(012) peak and (018)/(110) peak can be clearly distinguished, indicating that the material is a typical layered structure [41]. The lattice parameters of the prepared samples were analyzed via JADE6.0, as shown in Table 1. Compared with SLS and SLC, c and c/a values increased in OCP, indicating that the crystal lattice preferentially grows along the *c* axis, thereby promoting the electrochemical reaction.

The SEM micrographs and elemental compositions of samples SLC, SLS, and OCP are shown in Figure 2. As shown in Figure 2a,b,d,e, the primary particles of the samples SLC and SLS have a size of about 400–600 nm. The agglomerated secondary particles have no obvious morphology. As shown in Figure 2g, the primary particle size of the sample OCP is about 300–500 nm. As can be seen from Figure 2h, the secondary particles are spheroidal and have a particle size of about

20 µm. The embedded image in Figure 2h shows the SEM image of OCP secondary particles at low magnification. In addition, there are pores on the surface of the secondary particles closely clustered by the primary particles. Figure 2c,f,i shows the energy dispersion spectra (EDS) of SLC, SLS, and OCP, respectively. All peaks correspond to the characteristic peaks of the O, Mn, Co, and Ni elements, and the ratios of the elements are shown in the interpolation table. It can be clearly seen that there are differences in the element ratios of the three samples, especially the Ni and Mn. This may be the effect of ammonium radicals because in the reaction system, ammonium radicals complex with metal ions, which affects the chelation reaction of citric acid and metal ions and the precipitation reaction of oxalate and metal ions. In addition, the pH value may also affect the elemental composition of the product and the uniformity of the element distribution. These results indicate that a lithium-rich layered oxide positive electrode material with a high nickel content has been acquired.

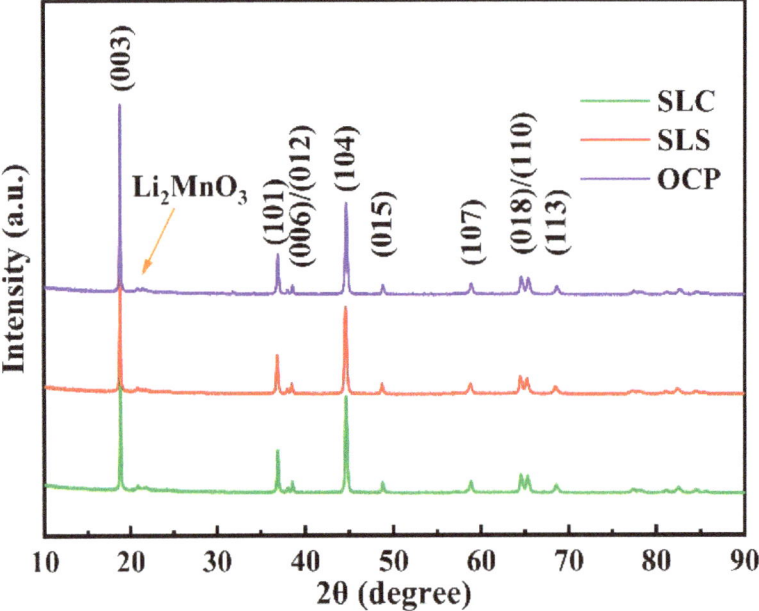

Figure 1. XRD patterns of LL-811 samples prepared by different methods.

Table 1. Lattice parameters of LL-811 samples prepared by different methods.

Sample	Lattice Parameters			c/a
	a (Å)	c (Å)	v (Å³)	
SLC	2.8564	14.2203	100.49	4.978
SLS	2.8605	14.2348	100.87	4.976
OCP	2.8551	14.2376	100.37	4.980

To further understand the effect of preparation methods on crystal microstructure, the microscopic morphology of the samples SLC, SLS, and OCP are shown in Figure 3. Figure 3a,e,i shows TEM images of SLC, SLS, and OCP samples, respectively, and the particle size is consistent with that of the SEM image. Figure 3b,f,j shows the HRTEM images of the corresponding area in Figure 3a,e,i. There is a clear lattice fringe in the "c" region of Figure 3b with interplanar spacing of 0.477 nm, corresponding to the (003) plane of the layered phase. These corresponding diffraction points of the (003) and (006) panel can be found in Figure 3c of its fast Fourier transform (FFT). The FFT of the "d" region in

Figure 3b is shown in Figure 3d, and there are diffraction points of the α-NaFeO$_2$ layered structure of the space group R$\bar{3}$m and the monoclinic system of the space group C2/m. Figure 3f shows very clear lattice fringes with a lattice spacing of 0.474 nm and 0.236 nm, corresponding to the (003) and (006) plane of the α-NaFeO$_2$ layered structure, respectively. The FFT pattern and inverse fast Fourier transform (IFFT) pattern of the "g" region are shown in Figure 3g,h. The lattice spacing measured in Figure 3j is 0.478 nm, corresponding to the (003) plane in the layered structure. Figure 3k,l shows the FFT pattern and IFFT pattern of the "k" region, respectively. These isolated two sets of lattice pattern of LiMO$_2$ and Li$_2$MnO$_3$ indicate lithium-rich layered oxides have composite structure in SLC and SLS samples, while the only one set of lattice pattern gives proof of solid solution in the OCP sample. Moreover, for OCP samples, the (003) plane has been significantly expanded compared to the other two samples, and it helps to improve the rate capacity of lithium-rich layered oxides. X-ray line scan element distribution (EDX) maps of SLC, SLS, and OCP samples are shown in Figure 4. The results show that the elements of OCP are uniformly distributed, and no nickel segregation occurs [36,42]. Evenly distributed transition metal elements help to enhance the stability of the layered structure and suppress capacity and voltage decay [21,36].

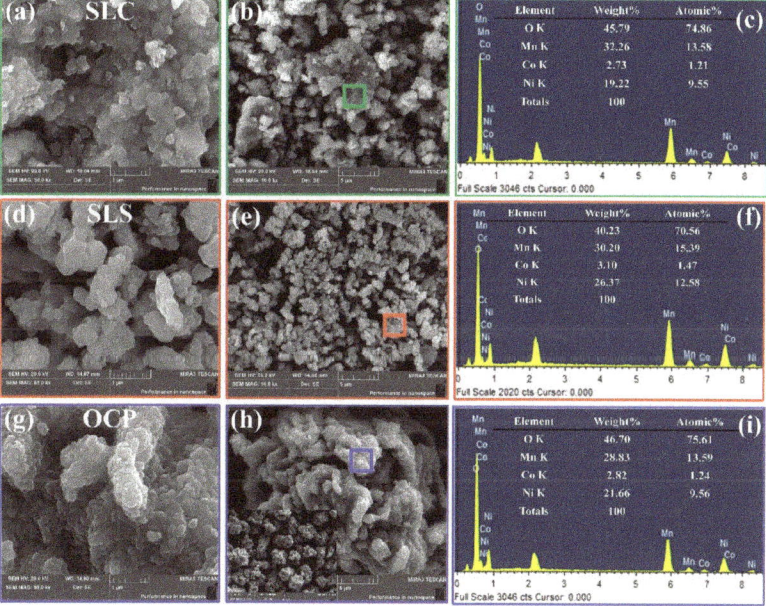

Figure 2. SEM images and energy dispersion spectra (EDS): (**a**–**c**) SLC; (**d**–**f**) SLS; (**g**–**i**) OCP.

The initial charge and discharge curves for all samples between 2.0 and 4.8 V at the current rate of 0.05 C are shown in Figure 5a. It shows similar initial charge and discharge curves, which is consistent with the characteristic curve of the lithium-rich layered oxide. The charging curve can be divided into an "S" zone below 4.5 V and an "L" zone above 4.5 V. The "S" region corresponds to the oxidation of the transition metal in the LiNi$_{0.8}$Co$_{0.1}$Mn$_{0.1}$O$_2$ component, and the "L" region corresponds to the Li and O removed from the crystal structure in the form of "Li$_2$O" [25]. As shown in Figure 5a, the initial discharge specific capacity of LL-811 prepared by the co-precipitation method is 262 mAh·g^{-1}, which is much higher than that of the SLC (220 mAh·g^{-1}) and SLS (231 mAh·g^{-1}) prepared by the sol–gel method. From the charging curve, the OCP sample has a longer 4.5 V platform compared to the SLC and SLS samples, which indicates that more non-electrochemically active Li$_2$MnO$_3$ is activated. According to the analysis of the TEM image (Figure 3), compared with the two-phase composite structure of SLC and SLS, the solid solution structure of OCP may be favorable for activating Li$_2$MnO$_3$. Figure 5b shows the initial charge–discharge curve of specific energy. OCP has the highest specific capacity, which is attributed to the fact that the lithium-rich layered oxide prepared by the oxalate co-precipitation method has smaller primary particles, and the secondary particles have a porous spherical morphology, which can shorten the diffusion pathway of Li$^+$ ions.

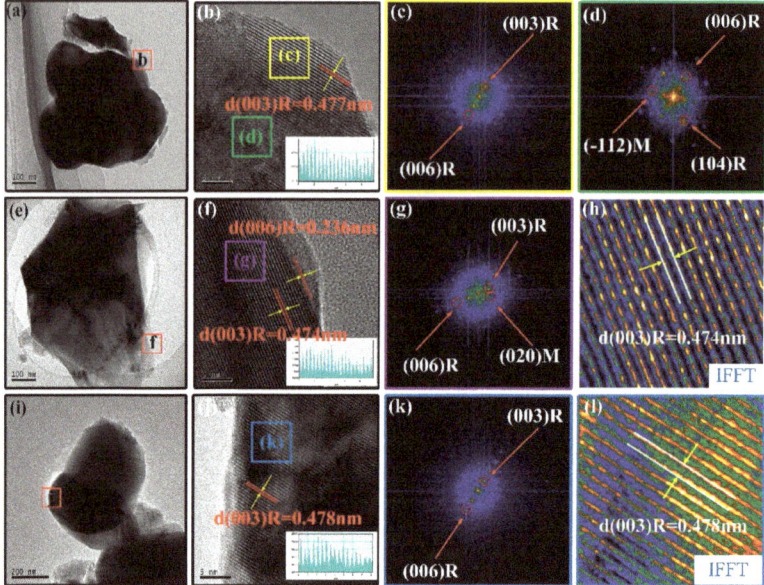

Figure 3. TEM and HRTEM images: (**a,b**) SLC; (**e,f**) SLS; (**i,j**) OCP; (**c,d**) fast Fourier transform (FFT) of the corresponding area in (**b**); (**g,h**) FFT and inverse fast Fourier transform (IFFT) of the corresponding area in (**f**); (**k,l**) FFT and IFFT of the corresponding area in (**j**). The indexes marked by R and M are related to the rhombohedral (R$\bar{3}$m) phase and monoclinic Li$_2$MnO$_3$ (C2/m) phase.

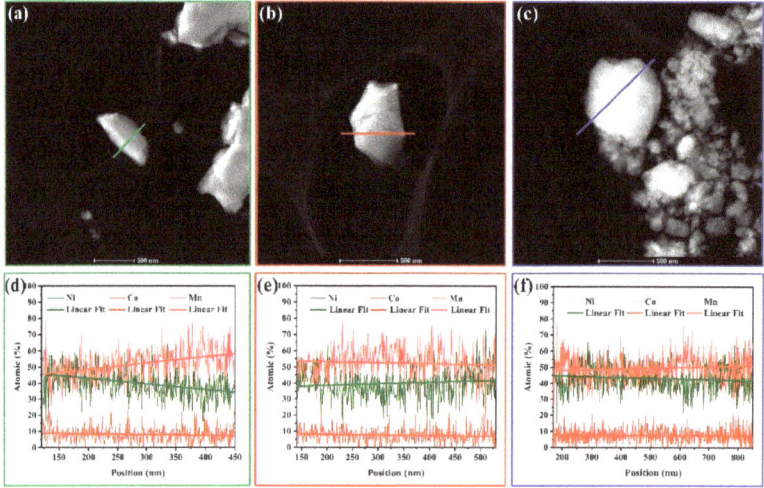

Figure 4. TEM images and X-ray line scan element distribution (EDX) maps: (**a,d**) SLC; (**b,e**) SLS; (**c,f**) OCP.

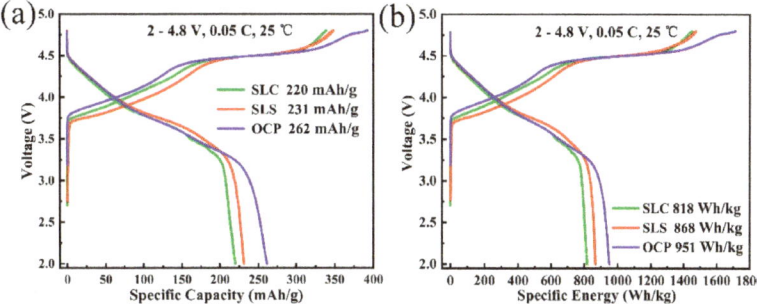

Figure 5. Initial charge and discharge curves of LL-811 samples prepared by different methods: (**a**) specific capacity; (**b**) specific energy.

Figure 6 shows the voltage and capacity decay of all samples between 2.0–4.6 V at a current rate of 1 C. As can be seen from Figure 6a, after 100 cycles, the discharge median voltage decay of OCP is 210 mV and the retention reaches 94.1%, which is higher than 91.8% of SLC and 87.9% of SLS, and the voltage decay is obviously suppressed. Figure 6b shows the specific capacity decay of the lithium-rich layered oxide cathode material. It can be seen that after 100 cycles, the specific capacity retention of OCP is 94.2%, which is significantly higher than 88.4% of SLC and 83.5% of SLS. By comparing the discharge specific energy of all samples, the results in Figure 6c are similar to those in Figure 6b. The discharge specific energy of OCP retention is 89.9%, while the SLC and SLS are only 81.2% and 74.2%, respectively. Figure 6d–f shows the discharge curves of SLC, SLS, and OCP at different cycle times. In the first 75 cycles, the specific capacities of the SLC and SLS were decayed from 166.9 mAh·g^{-1} to 152.1 mAh·g^{-1} and 179.5 mAh·g^{-1} to 152.2 mAh·g^{-1}, respectively. However, the specific capacity of OCP was only decayed from 182.3 mAh·g^{-1} to 178.7 mAh·g^{-1}, and the capacity decay is much smaller than SLC and SLS. For the discharge median voltage, the decay of OCP is also significantly suppressed. OCP has a lowest voltage decay and a larger specific capacity after 75 cycles, indicating that its layered structure is more stable. The voltage and capacity decay of the OCP samples was suppressed, which can be attributed to the high nickel content, uniform element distribution, and stable layered structure [25,26,36]. In addition, the secondary particle structure is tight, and the presence of pores on the surface is also the reason for the best electrochemical performance of OCP.

To further understand the effect of the preparation method on the rate capability, the rate capability of SLC, SLS, and OCP is shown in Figure 7. The samples prepared by oxalate co-precipitation show higher discharge capacity at various rates. The specific capacities of OCP are 261.6, 233.4, 200.9, 184.6, 157.4, and 138.0 mAh·g^{-1} at the discharge rates of 0.05, 0.1, 0.5, 1, 3, and 5 C, respectively. This indicates that increasing the spacing of the (003) plane, increasing the contact area of the electrolyte with the positive electrode material, and shortening the lithium ion diffusion path can increase the rate capacity of the lithium-rich layered oxide. In addition, the solid solution structure may also help increase the specific capacity of lithium-rich layered oxides. Figure 7b,c shows the discharge curves of SLC, SLS, and OCP at different discharge rates. As the rates increase, the discharge capacity and voltage of all samples have different degrees of decay. The results show that when the battery is discharged at a higher current density, the electrode resistance increases significantly, and the discharge energy is greatly reduced, which seriously affects the application of the lithium-rich layered oxide positive electrode material in electric vehicles. Although the OCP sample has the highest specific capacity under different current densities, its specific capacity has a greater decay at large current densities, resulting in a lower rate capacity retention than the SLS sample.

In order to understand the influence of the preparation method on the interfacial electrochemical and reaction kinetics of LL-811, the electrochemical impedance spectroscopy (EIS) of LL-811 prepared by different preparation methods were investigated. The Nyquist plots of SLC, SLS, and OCP are shown in Figure 8a. The impedance spectrum was fitted using the embedded equivalent circuit in Figure 8a, and the fitting results are shown in Table 2. The results show that the SLS has the smallest Rf and OCP has the smallest charge-transfer resistance (Rct). Figure 8b shows the linear relationship between Z' and $\omega^{-1/2}$, and the slope obtained by linear fitting represents the value of σ. The lithium ion diffusion coefficients calculated by the formula are shown in Table 2 [43]. The lithium ion diffusion coefficient of OCP is 3.67×10^{-13} cm^2·s^{-1}, which is higher than 2.04×10^{-13} cm^2·s^{-1} of SLC and 1.86×10^{-13} cm^2·s^{-1} of SLS. This proves that the lithium-rich layered oxide prepared by the oxalic acid co-precipitation method has a faster migration rate of lithium ions, and the rate performance is excellent in comparison.

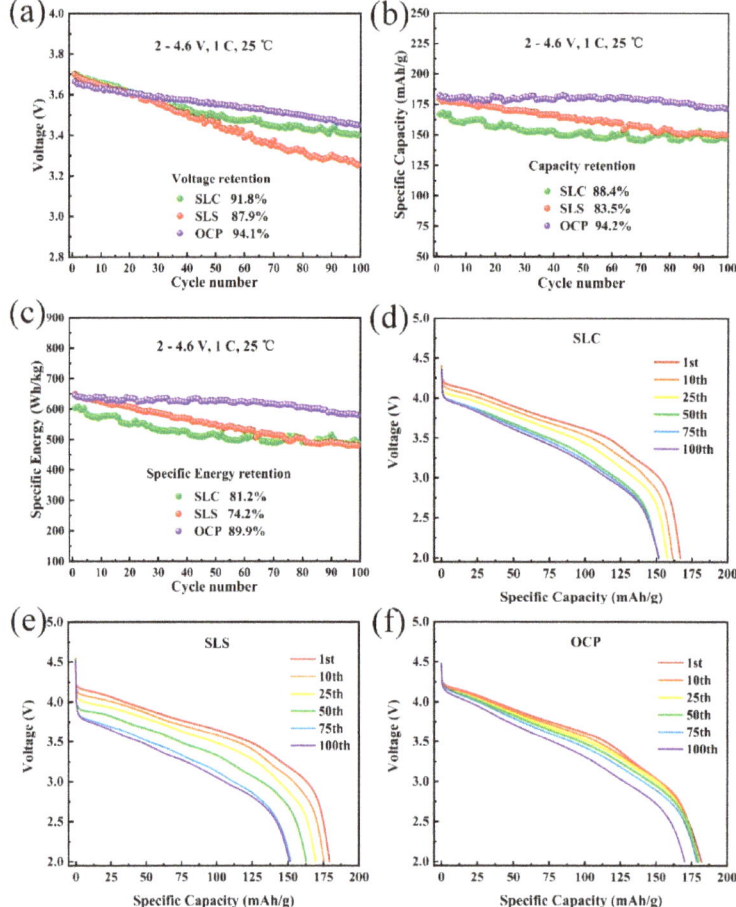

Figure 6. Voltage and capacity decay of LL-811 samples prepared by different methods: (**a**) voltage decay, (**b**) specific capacity decay, (**c**) specific energy decay, (**d–f**) discharge curves for different cycles.

Table 2. Impedance and lithium ion diffusion coefficient of LL-811 prepared by different methods.

Sample	R_f (Ω)	Rct (Ω)	D_{Li}^+ (cm^2·s^{-1})
SLC	230	503	2.04×10^{-13}
SLS	173	491	1.86×10^{-13}
OCP	246	369	3.67×10^{-13}

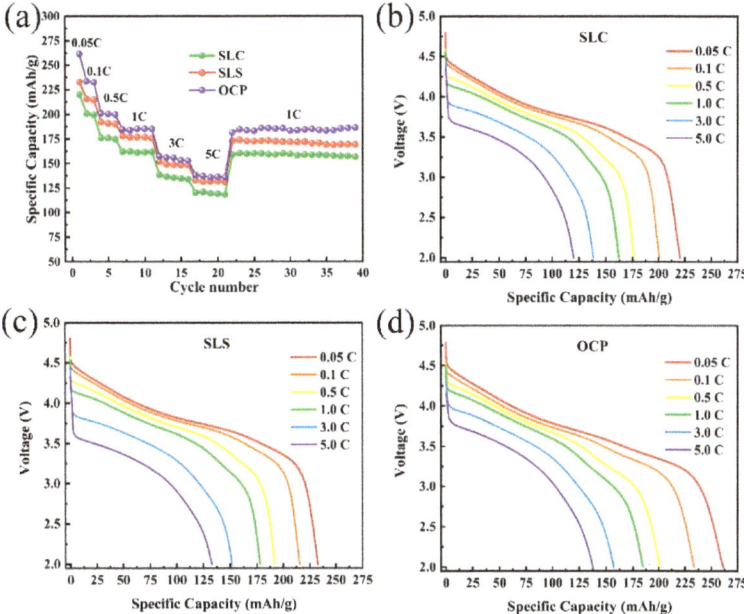

Figure 7. Rate performance of LL-811 samples prepared by different methods: (**a**) rate capacity, (**b**–**d**) discharge curves at different rates.

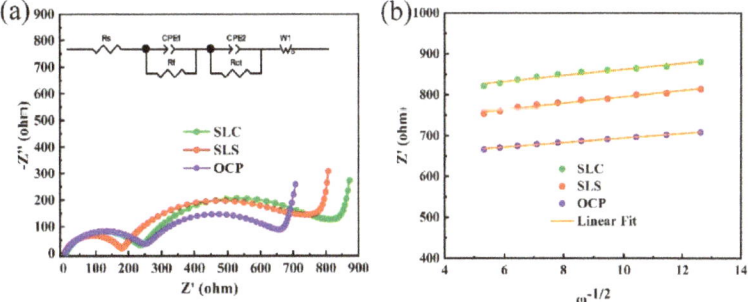

Figure 8. (**a**) Nyquist plots and (**b**) plots comparison of Z' vs $\omega^{-1/2}$ for LL-811 samples after the first cycle.

4. Conclusions

The effects of preparation methods on the structure, morphology, and electrochemical properties of $0.5Li_2MnO_3 \cdot 0.5LiMn_{0.8}Ni_{0.1}Co_{0.1}O_2$ cathode materials were systematically investigated. The results show that the lithium-rich layered oxide prepared by the oxalate co-precipitation method has the best performance. After 100 cycles at 1 C, the voltage and capacity decayed were only 210 mV and 10 mAh·g^{-1}, and the retention rates were 94.1% and 94.2%, respectively. The specific capacities of OCP are 261.6, 233.4, 200.9, 184.6, 157.4, and 138.0 mAh·g^{-1} at the discharge rates of 0.05, 0.1, 0.5, 1, 3, and 5 C, respectively. The significant reduction in voltage decay can be attributed to the high nickel content and uniform element distribution. In addition, tightly packed porous spheres help to reduce lithium ion diffusion energy and improve cycle stability and rate capacity. Therefore, the synthesis method plays an important role in the preparation of high-energy-density lithium-rich layered oxide cathode materials. This conclusion provides a reference for designing high-energy-density lithium-ion batteries.

Author Contributions: Conceptualization and methodology, Z.C. and Q.L.; investigation, J.L. and C.Z.; resources, X.L., Z.T. and M.Y.; Writing—Original draft preparation, Q.L.; Writing—Review and editing, Z.C.; supervision, J.D. and L.L.; project administration, H.Z. All authors have read and agreed to the published version of the manuscript.

Funding: This work was supported by the National Natural Science Foundation of China (No. 51874048), the Research Foundation of Education Bureau of Hunan Province (No. 19A003) and Scientific Research Fund of Changsha Science and Technology Bureau (No. kq1901100) and Postgraduate Innovative Test Program of Hunan Province.

Conflicts of Interest: The authors declare no conflict of interest.

References

1. Cano, Z.P.; Banham, D.; Ye, S.Y.; Hintennach, A.; Lu, J.; Fowler, M.; Chen, Z.W. Batteries and fuel cells for emerging electric vehicle markets. *Nat. Energy* **2018**, *3*, 279–289. [CrossRef]
2. Chu, S.; Cui, Y.; Liu, N. The path towards sustainable energy. *Nat. Mater.* **2016**, *16*, 16–22. [CrossRef] [PubMed]
3. Zhu, C.; Wei, D.H.; Wu, Y.L.; Zhang, Z.; Zhang, G.H.; Duan, J.F.; Li, L.J.; Zhu, H.L.; Zhu, Z.Y.; Chen, Z.Y. Controllable construction of interconnected SnO/N-doped carbon/carbon composite for enhanced-performance lithium-ion batteries anodes. *J. Alloy. Compd.* **2019**, *778*, 731–740. [CrossRef]
4. Ji, Y.; Zhou, C.K.; Lin, F.; Li, B.J.; Yang, F.F.; Zhu, H.L.; Duan, J.F.; Chen, Z.Y. Submicron Sized Nb Doped Lithium Garnet for High Ionic Conductivity Solid Electrolyte and Performance of All Solid-State Lithium Battery. *Preprints* **2019**, 2019120307. [CrossRef]
5. Goodenough, J.B.; Kim, Y. Challenges for Rechargeable Li Batteries. *Chem. Mater. Rev.* **2010**, *22*, 587–603. [CrossRef]
6. Nitta, N.; Wu, F.X.; Lee, J.T.; Yushin, G. Li-ion battery materials: Present and future. *Mater. Today* **2015**, *18*, 252–264. [CrossRef]
7. Schmuch, R.; Wagner, R.; Hörpel, G.; Placke, T.; Winter, M. Performance and cost of materials for lithium-based rechargeable automotive batteries. *Nat. Energy* **2018**, *3*, 267–278. [CrossRef]
8. Andre, D.; Kim, S.J.; Peter, L.; Lux, S.F.; Maglia, F.; Paschosa, O.; Stiasznya, B. Future generations of cathode materials: An automotive industry perspective. *J. Mater. Chem. A* **2015**, *3*, 6709–6732. [CrossRef]
9. Manthiram, A.; Song, B.H.; Li, W.D. A perspective on nickel-rich layered oxide cathodes for lithium-ion batteries. *Energy Storage Mater.* **2017**, *6*, 125–139. [CrossRef]
10. Myung, S.T.; Maglia, F.; Park, K.J.; Yoon, C.S.; Lamp, P.; Kim, S.J.; Sun, Y.K. Nickel-Rich Layered Cathode Materials for Automotive Lithium-Ion Batteries: Achievements and Perspectives. *ACS Energy Lett.* **2017**, *2*, 196–223. [CrossRef]
11. Li, M.; Lu, J.; Chen, Z.W.; Amine, K. 30 Years of Lithium-Ion Batteries. *Adv. Mater.* **2018**, *30*, 1800561–1800584. [CrossRef] [PubMed]
12. Zhang, H.L.; Zhao, H.B.; Khan, M.A.; Zou, W.W.; Xu, J.Q.; Zhang, L.; Zhang, J.J. Recent progress in advanced electrode materials, separators and electrolytes for lithium batteries. *J. Mater. Chem. A* **2018**, *6*, 20564–20620. [CrossRef]
13. Liu, J.; Bao, Z.N.; Cui, Y.; Dufek, E.J.; Goodenough, J.B.; Khalifah, P.; Li, Q.Y.; Liaw, B.Y.; Liu, P.; Manthiram, A.; et al. Pathways for practical high-energy long-cycling lithium metal batteries. *Nat. Energy* **2019**, *4*, 180–186. [CrossRef]
14. Thackeray, M.M.; Kang, S.H.; Johnson, C.S.; Vaughey, J.T.; Benedek, R.; Hackney, S.A. Li$_2$MnO$_3$-stabilized LiMO$_2$ (M = Mn, Ni, Co) electrodes for lithium-ion batteries. *J. Mater. Chem.* **2007**, *17*, 3112–3125. [CrossRef]
15. Mohanty, D.; Li, J.L.; Abraham, D.P.; Huq, A.; Payzant, E.A.; Wood, D.L.; Daniel, C. Unraveling the Voltage-Fade Mechanism in High-Energy-Density Lithium-Ion Batteries: Origin of the Tetrahedral Cations for Spinel Conversion. *Chem. Mater.* **2014**, *26*, 6272–6280. [CrossRef]
16. Sathiya, M.; Abakumov, A.M.; Foix, D.; Rousse, G.; Ramesha, K.; Saubanere, M.; Doublet, M.L.; Vezin, H.; Laisa, C.P.; Prakash, A.S.; et al. Origin of voltage decay in high-capacity layered oxide electrodes. *Nat. Mater.* **2015**, *14*, 230–239. [CrossRef]

17. Hu, E.Y.; Yu, X.Q.; Lin, R.Q.; Bi, X.X.; Lu, J.; Bak, S.; Nam, K.W.; Xin, H.L.; Jaye, C.; Fischer, D.A.; et al. Evolution of redox couples in Li- and Mn-rich cathode materials and mitigation of voltage fade by reducing oxygen release. *Nat. Energy* **2018**, *3*, 690–698. [CrossRef]
18. Singer, A.; Zhang, M.; Hy, S.; Cela, D.; Fang, C.; Wynn, T.A.; Qiu, B.; Xia, Y.; Liu, Z.; Ulvestad, A.; et al. Nucleation of dislocations and their dynamics in layered oxide cathode materials during battery charging. *Nat. Energy* **2018**, *3*, 641–647. [CrossRef]
19. Zhu, H.L.; Li, Q.F.; Gong, X.L.; Cao, K.F.; Chen, Z.Y. Enhanced High Voltage Performance of Chlorine/Bromine Co-Doped Lithium Nickel Manganese Cobalt Oxide. *Crystals* **2018**, *8*, 425. [CrossRef]
20. Pang, W.K.; Lin, H.F.; Peterson, V.K.; Lu, C.Z.; Liu, C.E.; Liao, S.C.; Chen, J.M. Effects of Fluorine and Chromium Doping on the Performance of Lithium-Rich $Li_{1+x}MO_2$ (M = Ni, Mn, Co) Positive Electrodes. *Chem. Mater.* **2017**, *29*, 10299–10311. [CrossRef]
21. Liu, D.M.; Fan, X.J.; Li, Z.H.; Liu, T.; Ling, M.; Liu, Y.J.; Liang, C.D. A cation/anion co-doped $Li_{1.12}Na_{0.08}Ni_{0.2}Mn_{0.6}O_{1.95}F_{0.05}$ cathode for lithium ion batteries. *Nano Energy* **2019**, *58*, 786–796. [CrossRef]
22. Zheng, J.M.; Gu, M.; Xiao, J.; Polzin, B.J.; Yan, P.F.; Chen, X.L.; Wang, C.M.; Zhang, J.G. Functioning Mechanism of AlF_3 Coating on the Li- and Mn-Rich Cathode Materials. *Chem. Mater.* **2014**, *26*, 6320–6327. [CrossRef]
23. Xu, M.; Chen, Z.Y.; Li, L.J.; Zhu, H.L.; Zhao, Q.F.; Xu, L.; Peng, N.F.; Gong, L. Highly crystalline alumina surface coating from hydrolysis of aluminum isopropoxide on lithium-rich layered oxide. *J. Power Sources* **2015**, *281*, 444–454. [CrossRef]
24. Liu, Y.J.; Fan, X.J.; Zhang, Z.Q.; Wu, H.H.; Liu, D.M.; Dou, A.C.; Su, M.R.; Zhang, Q.B.; Chu, D.W. Enhanced electrochemical performance of Li-rich layered cathode materials by combined Cr doping and $LiAlO_2$ coating. *ACS Sustain. Chem. Eng.* **2019**, *7*, 2225–2235. [CrossRef]
25. Shi, J.L.; Xiao, D.D.; Ge, M.Y.; Yu, X.Q.; Chu, Y.; Huang, X.J.; Zhang, X.D.; Yin, Y.X.; Yang, X.Q.; Guo, Y.G.; et al. High-Capacity Cathode Material with High Voltage for Li-Ion Batteries. *Adv. Mater.* **2018**, *30*, 1705575–1705582. [CrossRef]
26. Shi, J.L.; Zhang, J.N.; He, M.; Zhang, X.D.; Yin, Y.X.; Li, H.; Guo, Y.G.; Gu, L.; Li, J.W. Mitigating Voltage Decay of Li-Rich Cathode Material via Increasing Ni-Content for Lithium-Ion Batteries. *ACS Appl. Mater. Interfaces* **2016**, *8*, 20138–20146. [CrossRef]
27. Ju, X.K.; Hou, X.; Liu, Z.Q.; Zheng, H.F.; Huang, H.; Qu, B.H.; Wang, T.H.; Li, Q.H.; Li, J. The full gradient design in Li-rich cathode for high performance lithium ion batteries with reduced voltage decay. *J. Power Sources* **2019**, *437*, 226902. [CrossRef]
28. Du, C.Q.; Zhang, F.; Ma, C.X.; Wu, J.W.; Tang, Z.Y.; Zhang, X.H.; Qu, D.Y. Synthesis and electrochemical properties of $Li_{1.2}Mn_{0.54}Ni_{0.13}Co_{0.13}O_2$ cathode material for lithium-ion battery. *Ionics* **2016**, *22*, 209–218. [CrossRef]
29. Tang, T.; Zhang, H.L. Synthesis and electrochemical performance of lithium-rich cathode material $Li[Li_{0.2}Ni_{0.15}Mn_{0.55}Co_{0.1-x}Al_x]O_2$. *Electrochim. Acta* **2016**, *191*, 263–269. [CrossRef]
30. Fu, F.; Huang, Y.Y.; Wu, P.; Bu, Y.K.; Wang, Y.B.; Yao, J.N. Controlled synthesis of lithium-rich layered $Li_{1.2}Mn_{0.56}Ni_{0.12}Co_{0.12}O_2$ oxide with tunable morphology and structure as cathode material for lithium-ion batteries by solvo/hydrothermal methods. *J. Alloy. Compd.* **2015**, *618*, 673–678. [CrossRef]
31. Shi, S.J.; Tu, J.P.; Tang, Y.Y.; Yu, Y.X.; Zhang, Y.Q.; Wang, X.L. Synthesis and electrochemical performance of $Li_{1.131}Mn_{0.504}Ni_{0.243}Co_{0.122}O_2$ cathode materials for lithium ion batteries via freeze drying. *J. Power Sources* **2013**, *221*, 300–307. [CrossRef]
32. Chen, Z.Y.; Yan, X.Y.; Xu, M.; Cao, K.F.; Zhu, H.L.; Li, L.J.; Duan, J.F. Building Honeycomb-Like Hollow Microsphere Architecture in a Bubble Template Reaction for High-Performance Lithium-Rich Layered Oxide Cathode Materials. *ACS Appl. Mater. Interfaces* **2017**, *9*, 30617–30625. [CrossRef] [PubMed]
33. Bareno, J.; Lei, C.H.; Wen, J.G.; Kang, S.H.; Petrov, I.; Abraham, D.P. Local structure of layered oxide electrode materials for lithium-ion batteries. *Adv. Mater.* **2010**, *22*, 1122–1127. [CrossRef] [PubMed]
34. Jarvis, K.A.; Deng, Z.Q.; Allard, L.F.; Manthiram, A.; Ferreira, P.J. Atomic Structure of a Lithium-Rich Layered Oxide Material for Lithium-Ion Batteries: Evidence of a Solid Solution. *Chem. Mater.* **2011**, *23*, 3614–3621. [CrossRef]
35. Long, B.R.; Croy, J.R.; Dogan, F.; Suchomel, M.R.; Key, B.; Wen, J.G.; Miller, D.J.; Thackeray, M.M.; Balasubramanian, M. Effect of Cooling Rates on Phase Separation in $0.5Li_2MnO_3 \cdot 0.5LiCoO_2$ Electrode Materials for Li-Ion Batteries. *Chem. Mater.* **2014**, *26*, 3565–3572. [CrossRef]

36. Zheng, J.M.; Gu, M.; Genc, A.; Xiao, J.; Xu, P.H.; Chen, X.L.; Zhu, Z.H.; Zhao, W.B.; Pullan, L.; Wang, C.M.; et al. Mitigating Voltage Fade in Cathode Materials by Improving the Atomic Level Uniformity of Elemental Distribution. *Nano Lett.* **2014**, *14*, 2628–2635. [CrossRef]
37. Shojan, J.; Rao, C.V.; Torres, L.; Singh, G.; Katiyar, R.S. Lithium-ion battery performance of layered $0.3Li_2MnO_3$-$0.7LiNi_{0.5}Mn_{0.5}O_2$ composite cathode prepared by co-precipitation and sol-gel methods. *Mater. Lett.* **2013**, *104*, 57–60. [CrossRef]
38. Wang, C.C.; Jarvis, K.A.; Ferreira, P.J.; Manthiram, A. Effect of Synthesis Conditions on the First Charge and Reversible Capacities of Lithium-Rich Layered Oxide Cathodes. *Chem. Mater.* **2013**, *25*, 3267–3275. [CrossRef]
39. Li, L.J.; Xu, M.; Chen, Z.Y.; Zhou, X.; Zhang, Q.B.; Zhu, H.L.; Wu, C.; Zhang, K.L. High-performance lithium-rich layered oxide materials: Effects of chelating agents on microstructure and electrochemical properties. *Electrochim. Acta* **2015**, *174*, 446–455. [CrossRef]
40. Liu, J.; Liu, Q.M.; Zhu, H.L.; Lin, F.; Ji, Y.; Li, B.J.; Duan, J.F.; Chen, Z.Y. Effect of Different Composition on Voltage Attenuation of Li-Rich Cathode Material for LithiumIon Batteries. *Materials* **2020**, *13*, 40. [CrossRef]
41. Yu, R.Z.; Zhang, Z.J.; Jamil, S.; Chen, J.C.; Zhang, X.H.; Wang, X.Y.; Yang, Z.H.; Shu, H.B.; Yang, X.K. Effects of Nanofiber Architecture and Antimony Doping on the Performance of Lithium-Rich Layered Oxides: Enhancing Lithium Diffusivity and Lattice Oxygen Stability. *ACS Appl. Mater. Interfaces* **2018**, *10*, 16561–16571. [CrossRef] [PubMed]
42. Gu, M.; Belharouak, I.; Genc, A.; Wang, Z.; Wang, D.; Amine, K.; Gao, F.; Zhou, G.; Thevuthasan, S.; Baer, D.R.; et al. Conflicting roles of nickel in controlling cathode performance in lithium ion batteries. *Nano Lett.* **2012**, *12*, 5186–5191. [CrossRef] [PubMed]
43. Chen, Z.Y.; Zhang, Z.; Zhao, Q.F.; Duan, J.F.; Zhu, H.L. Understanding the Impact of K-Doping on the Structure and Performance of LiFePO$_4$/C Cathode Materials. *J. Nanosci. Nanotechnol.* **2019**, *19*, 119–124. [CrossRef] [PubMed]

 © 2020 by the authors. Licensee MDPI, Basel, Switzerland. This article is an open access article distributed under the terms and conditions of the Creative Commons Attribution (CC BY) license (http://creativecommons.org/licenses/by/4.0/).

Article

Enhanced Electrochemical Performances of Cobalt-Doped Li$_2$MoO$_3$ Cathode Materials

Zhiyong Yu [1,2,*], Jishen Hao [1,2], Wenji Li [1,2] and Hanxing Liu [1,3,*]

[1] State Key Laboratory of Advanced Technology for Materials Synthesis and Processing, Wuhan University of Technology, Wuhan 430070, China; haojishen@live.com (J.H.); lwjwhut@126.com (W.L.)
[2] School of Materials Science and Engineering, Wuhan University of Technology, Wuhan 430070, China
[3] International School of Materials Science and Engineering, Wuhan University of Technology, Wuhan 430070, China
* Correspondence: yuzhiyong@whut.edu.cn (Z.Y.); lhxhp@whut.edu.cn (H.L.)

Received: 7 February 2019; Accepted: 9 March 2019; Published: 13 March 2019

Abstract: Co-doped Li$_2$MoO$_3$ was successfully synthesized via a solid phase method. The impacts of Co-doping on Li$_2$MoO$_3$ have been analyzed by X-ray photoelectron spectroscopy (XPS), X-ray powder diffraction (XRD), scanning electron microscope (SEM), and Fourier transform infrared spectroscopy (FTIR) measurements. The results show that an appropriate amount of Co ions can be introduced into the Li$_2$MoO$_3$ lattices, and they can reduce the particle sizes of the cathode materials. Electrochemical tests reveal that Co-doping can significantly improve the electrochemical performances of the Li$_2$MoO$_3$ materials. Li$_2$Mo$_{0.90}$Co$_{0.10}$O$_3$ presents a first-discharge capacity of 220 mAh·g^{-1}, with a capacity retention of 63.6% after 50 cycles at 5 mA·g^{-1}, which is much better than the pristine samples (181 mAh·g^{-1}, 47.5%). The enhanced electrochemical performances could be due to the enhancement of the structural stability, and the reduction in impedance, due to the Co-doping.

Keywords: Li$_2$MoO$_3$; Co-doping; cathode materials; Li ion battery

---

1. Introduction

Recently, the development of high-capacity cathode materials has become a hot topic in the field of Li-ion batteries. Mn-based Li-rich layer oxides xLi$_2$MnO$_3$·(1 − x)LiMO$_2$ (0 < x < 1.0, M = Mn, Ni, Co, etc.) were proposed as potential cathode materials, due to their high discharge capacities of above 280 mAh·g^{-1}, and thus, the structure stability of the Li$_2$MnO$_3$ component [1–8]. Unfortunately, numerous reports have indicated that the drawbacks of Li$_2$MnO$_3$-based composites, such as low initial Coulombic efficiency, a fast decline in capacity, and potential safety hazards, were difficult to overcome, which severely restricted their practice applications [9–11]. Thus, much attention has been paid to find other transition metals instead of Mn, to build new Li$_2$MO$_3$ (M = Ru, Ir, Mo, etc.)-based materials for next generation Li–ion batteries in recent years [12–15].

Li$_2$MoO$_3$ as a type of Li–rich layer cathode material with alternating Li layers and randomly distributed [Li$_{1/3}$Mo$_{2/3}$] layers, has attracted much research interest [15–21]. The previous studies verified that Li$_2$MoO$_3$ promised a high theoretical capacity of up to 339 mAh·g^{-1}, and a near-absence of oxygen evolution [17,18], which supported Li$_2$MoO$_3$ as a candidate to replace Li$_2$MnO$_3$ in constructing Li-rich cathode materials. However, the poor cycling stability and rate capability of the Li$_2$MoO$_3$ material, owing to its low conductivity and irreversible phase transition, hinders its practical application. Hence, it is necessary to find a suitable modification method to improve the performance of the Li$_2$MoO$_3$ material.

At present, only a few studies about on modifying Li$_2$MoO$_3$ have been reported [19–21]. Ceder's group constructed a solid solution between Li$_2$MoO$_3$ and LiCrO$_2$ for cathode materials [19]. The Li$_2$MoO$_3$–LiCrO$_2$ cathode materials presented not only high-discharge capacities, but also great cycling stabilities over the 10 cycles. In our previous study, carbon-coated Li_2MoO_3 composites were successfully prepared, and they achieved much lower impedances and better electrochemical performances than bare Li_2MoO_3 [21]. Cobalt doping has been considered to be a facile and effective method in enhancing the electrochemical performances, since it can improve structure stability and reduce the impedance of cathode materials [22–26]. In this paper, cobalt was selected to improve the electrochemical performances of Li_2MoO_3 for the first time. The structural characteristics and electrochemical performances of $Li_2Mo_{1-x}Co_xO_3$ are presented here.

2. Materials and Methods

2.1. Preparation of the $Li_2Mo_{1-x}Co_xO_3$ Powder

The pristine and Co-doped Li_2MoO_3 powders were synthesized via a solid reaction method, as shown in Figure 1. Firstly, stoichiometric amounts of Li_2CO_3 (>99.7%, Sinopahrm Medicine, Shanghai, China), MoO_3 (>99.5%, Aldrich, Shanghai, China), and $2CoCO_3·3Co(OH)_2$ (>99.5%, Aldrich, Shanghai, China) were homogeneously mixed by ball milling, and then calcinated at 873 K for 24 h under air, to obtain the precursor. Li_2CO_3 was added in at 10% excess to compensate Li volatilization. After that, the obtained precursor was reduced in a stream of flowing $5\%H_2/95\%N_2$ at 973 K for 48 h to prepare $Li_2Mo_{1-x}Co_xO_3$.

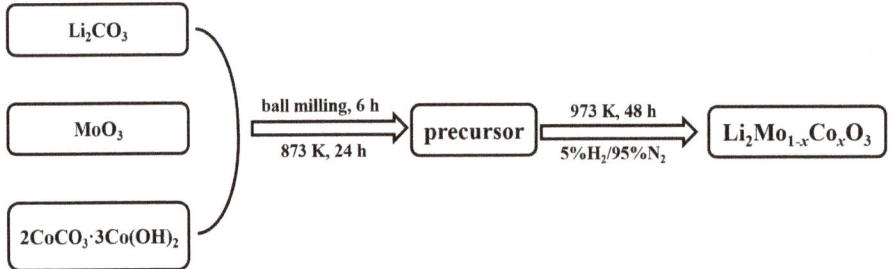

Figure 1. Flow chart of the process for preparing $Li_2Mo_{1-x}Co_xO_3$.

2.2. Physical Characterization

X-ray powder diffraction (XRD) was carried out by using a PhilipsX' Pert PW3050/60 diffractometer (PANalytical. B. V, Lelyweg, the Netherlands), with a scan rate of 0.02° per second, by Cu–Kα radiation (λ = 1.5406 Å). X'pert Highscor software (PANalytical. B. V, Lelyweg, the Netherlands) was used for Rietveld refinement. The morphologies of the samples was determined by a HITACHI S-4800 field-emission high-resolution scanning electron microscope (SEM) (Hitachi, Tokyo, Japan). X-ray photoelectron spectroscopy (XPS) was detected on a ESCALAB250Xi (ThermoFisher, Waltham, America). Fourier transform infrared spectroscopy (FTIR) was detected on Nicolet6700 (ThermoFisher, Waltham, America) in the wave range of 4000–400 cm^{-1} with a high resolution of 4 cm^{-1}.

2.3. Electrochemical Tests

The electrochemical performances were tested through CR2032-type coin cells (HF-Kejing, Hefei, China). Cathodes were prepared by mixing 70 wt % $Li_2Mo_{1-x}Co_xO_3$, 20 wt % acetylene carbon black and 10 wt % polyvinylidene fluoride (PVDF) in N-methyl-2-pyrrolidone (NMP) solution. The slurry was cast evenly onto a stainless steel sheet, and dried in a vacuum oven (Suopuyiqi, Shanghai, China) at 120 °C for 12 h. Lithium metal and Celgard 2400 were used as the anode and separator, respectively. A concentration of 1 mol·L^{-1} $LiPF_6$ in ethylene carbonate (EC)/dimethyl carbonate (DMC) (volume ratio: 1:1) solution was adopted as the electrolyte. The coin cells were assembled in an argon-filled glove box, and measured on a Land CT2001A system (LANHE, Wuhan, China) in galvanostatic mode at 30 °C. Electrochemical impedance spectroscopy (EIS) was tested using an electrochemical station (CHI660B) (Chenhua, Shanghai, China) in 10^{-2} Hz–10 MHz, with a voltage amplitude of 5 mV. Cycliation at a scanning speed of 10^{-1} mVc voltammetry (CV) was performed with the same electrochemical st s^{-1}.

3. Results and Discussion

3.1. Characteristics of the as-Prepared $Li_2Mo_{1-x}Co_xO_3$

Figure 2a and b shows the XPS spectra of Co $2p_{3/2}$ in $Li_2Mo_{0.90}Co_{0.10}O_3$, and the corresponding precursor. The peak of Co $2p_{3/2}$ in the precursor is at 779.8 eV (Figure 2a) with a weak satellite at 789.9 eV. This satellite peak is 10.1 eV above Co $2p_{3/2}$, corresponding diamagnetic Co^{3+} (S = 0), which is very consistent with the XPS result reported for $LiCoO_2$ [27]. After the reduction by hydrogen at a high temperature, the original satellite disappears, and a new intense satellite appears at 784.4 eV, as shown in Figure 2b. Compared with the original satellite, the new satellite is 4.1 eV higher than the core level line (780.3 eV), corresponding to high-spin Co^{2+} (S = 3/2) compounds, indicating that the valence state of the Co element will be reduced from Co^{3+} to Co^{2+} after reduction processing under hydrogen [28]. Figure 2c and d show that the XPS spectra of Mo in Li_2MoO_3 and $Li_2Mo_{0.90}Co_{0.10}O_3$. The peak at around 230.1 eV is assigned to Mo^{4+} $3d_{5/2}$, and the peak at around 232.6 eV is assigned to Mo^{6+} $3d_{5/2}$. Obviously, after cobalt doping, the peak intensity of Mo^{6+} $3d_{5/2}$ rises, and the peak intensity of Mo^{4+} $3d_{5/2}$ decreases, indicating an increased amount of Mo^{6+}.

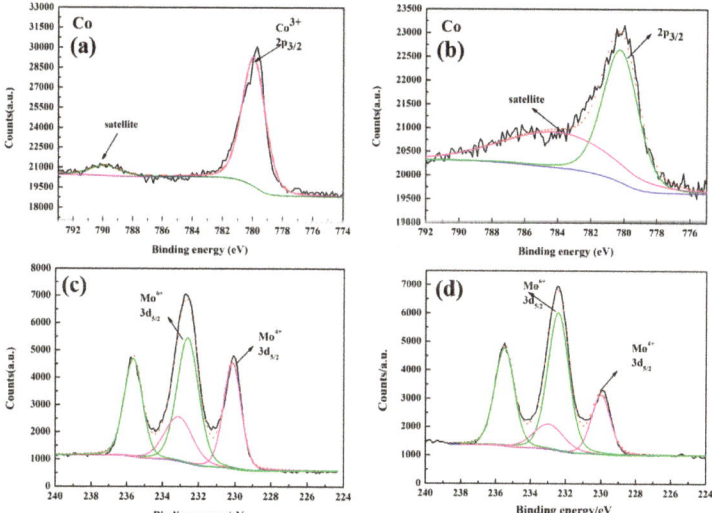

Figure 2. XPS spectra of Co in (**a**) the doped precursor, and (**b**) $Li_2Mo_{0.90}Co_{0.10}O_3$ and the XPS spectra of Mo in (**c**) Li_2MoO_3 and (**d**) $Li_2Mo_{0.90}Co_{0.10}O_3$.

Figure 3a shows the XRD patterns of the synthesized $Li_2Mo_{1-x}Co_xO_3$ (x = 0, 0.05, 0.10, 0.15). Except the sample with x = 0.15, all samples match well with the α-$NaFeO_2$ structure, which could be indexed to Li_2MoO_3 (see Figure 3a). When the Co content adds up to 0.15, the characteristic peaks of the impurity phase Li_4MoO_5 and Co appear. The splitting of the (006)/(101) peaks at 36°, reflecting that the layer structure weakens with the increase of Co content, indicating that Co-doping increases the disorder of the cations. Rietveld refinements for $Li_2Mo_{1-x}Co_xO_3$ were carried out, to obtain more information from XRD (see Figure 3b,c). The a(b)-parameters are increased, and c-parameters are decreased, with a rise of Co-doping content. Notably, the variations of cell parameters are negligible while x is higher than 0.10 considering the fitting error, which indicates that the solubility limit of Co is around $x = 0.10$. Moreover, the values of c/a drop with the increase of the Co-doping content, which indicates that Co-doping increases the disorder of Li_2MoO_3 materials. As indicated in the above XPS results, the valence of cobalt should be +2 in the samples. Its radius (0.745 Å) is very similar to that of Li^+ (0.76 Å), which may result in the increase of disorder. Figure 3d exhibits the unit cell volume of the pristine and the Co-doped Li_2MoO_3. Clearly, the unit cell volume increases with the rise of the Co-doping contents, which could be related to the replacement of Mo^{4+} (0.65 Å) by Co^{2+} (0.745 Å).

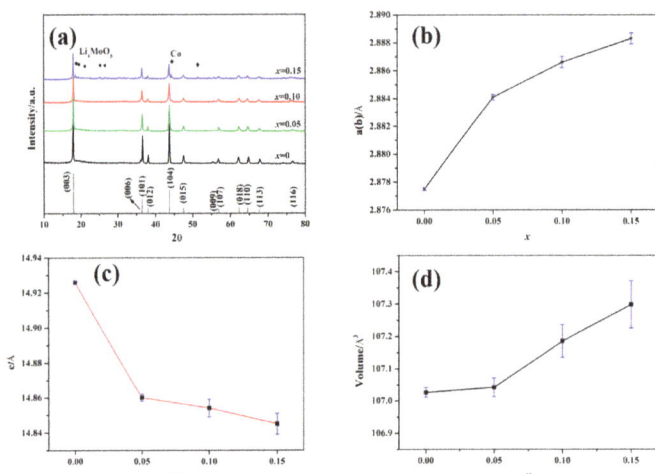

Figure 3. (a) XRD patterns, (b) a(b)-parameters in the lattice, (c) c-parameters, and (d) unit cell volume of the synthesized $Li_2Mo_{1-x}Co_xO_3$ ($x = 0, 0.05, 0.10, 0.15$).

In order to observe the impacts of Co-doping on the particle morphologies of the samples, the SEM images of $Li_2Mo_{1-x}Co_xO_3$ ($x = 0, 0.05, 0.10$) were examined (see Figure 4). The particle sizes of the pristine Li_2MoO_3 present a wide distribution range from 1 to 3 μm (see Figure 4a), whereas the doped samples show smaller particles with more uniform distributions in the range of 200–300 nm (see Figure 4b,c). The results suggest that the addition of Co affects the morphology, and decreases the particle size of cathode materials. Particle growth may be restricted by lattice distortion of Li_2MoO_3 due to the replacement of Mo by Co. Similar phenomena have also been observed by some other groups [29,30]. Ma, J. et al. studied the stability of Li_2MoO_3 in air [16]. Their results verified that Li_2MoO_3 easily adsorbed O_2 and thus was partially oxidized to Li_2MoO_4. Meanwhile, the CO_2 in air also reacted with Li_2MoO_3 to produce Li_2CO_3, which consumed the Li ions near the surface and produced MoO_3 [16]. In order to investigate the effects of Co-doping on the stability of $Li_2Mo_{1-x}Co_xO_3$ in air, FTIR spectra of Li_2MoO_3 and $Li_2Mo_{0.90}Co_{0.10}O_3$ were carried out (see Figure 5). The samples were stored in the air for 7 days before the FTIR test. Obviously, both samples present a similar FTIR spectra. The peaks at 446, 497, 559 and 698 cm^{-1} are assigned to Li_2MoO_3, which is consistent with the previous study [16]. While the peaks at 1480, 1420, 830 and 817 cm^{-1} could be attributed to Li_2CO_3 and Li_2MoO_4, respectively. These species are believed to be the reaction products between of Li_2MoO_3, CO_2 and O_2, revealing that both samples are partially decomposed in air. No peaks related MoO_3 are detected in the FTIR spectra, which may contribute to the shorter storage time compared with the previous report [16].

Figure 4. The SEM images of (**a**) Li_2MoO_3, (**b**) $Li_2Mo_{0.95}Co_{0.05}O_3$ and (**c**) $Li_2Mo_{0.90}Co_{0.10}O_3$.

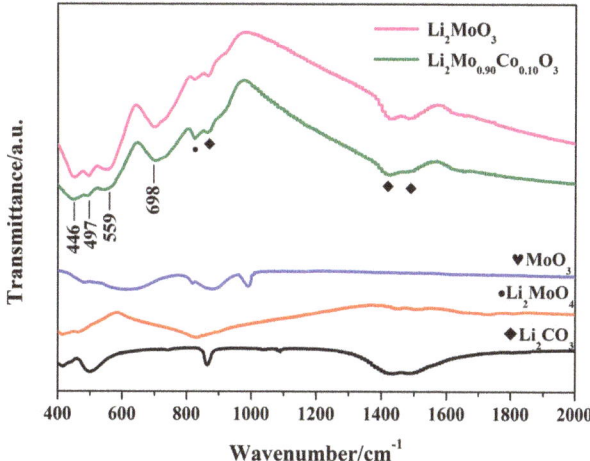

Figure 5. The FTIR spectra of Li_2MoO_3 and $Li_2Mo_{0.90}Co_{0.10}O_3$.

3.2. Electrochemical Performances

Figure 6 compares the initial charge-discharge profiles of pristine and Co-doped samples in the voltage range of 1.5–4.5 V at 5 mA·g^{-1}. The samples present initial discharge capacities of 181, 213, 220 and 165 mAh·g^{-1}, respectively. When the doping content is 0.05 and 0.10, the first discharge capacities of Co-doped samples are higher than that of the pristine sample. It also can be found the first discharge capacity decreases obviously while x = 0.15. What's more, voltage difference between charge and discharge profiles of $Li_2Mo_{0.90}Co_{0.10}O_3$ is much smaller than that of pristine Li_2MoO_3, indicating that Co-doping can effectively suppress the polarization and enhance reversibility of Li_2MoO_3 materials. Notable that the charge behaviors with two regions for Co-doped Li_2MoO_3 cathode materials are similar to that for the pristine sample in the first charge-discharge process, which may relate to the delithiation reaction corresponding to the oxidation of the Mo ions in the 1.5–3.7 V region and a Li_2MoO_3-$Li_{0.91}MoO_3$ two phase reaction in the 3.7–4.5 V region [17].

Figure 7 shows the cycling performances of pristine and Co-doped samples between 1.5 and 4.5 V at 5 mA·g^{-1}. It is clearly seen that $Li_2Mo_{0.95}Co_{0.05}O_3$ and $Li_2Mo_{0.90}Co_{0.10}O_3$ deliver much higher discharge capacity than that of pristine sample. After 50 cycles, pristine and Co-doped sample present the discharge capacities of 86, 121, 140 and 52 mAh·g^{-1} with the capacity retentions of 47.5%, 56.8%, 63.6% and 31.5%, respectively. Clearly, the cycling stability of pristine is poor. However, while increasing cobalt content to 0.05 and 0.10, the discharge capacity and cycling stability are significantly improved. With a further increase of Co content, the discharge capacity and cycling stability reduce, which may be attributed to the rise of inert impurities of Li_4MoO_5 and Co. Li_4MoO_5 delivers poor electrochemical performances because of its low electron conduction, low Coulombic efficiency and critical irreversible phase transition [31]. In addition, Co shows ignorable specific capacity above

1.5 V [32]. Therefore, the appearance of Co and Li_4MoO_5 in the sample has negative effects on the electrochemical performance of Li_2MoO_3. $Li_2Mo_{0.90}Co_{0.10}O_3$ possesses the highest discharge capacity and the best capacity retention, which indicates the amounts of dopant will be important for the electrochemical performances of Li_2MoO_3.

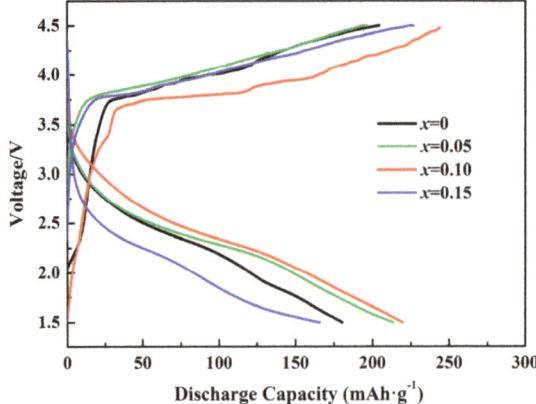

Figure 6. The initial charge–discharge profiles of the synthesized $Li_2Mo_{1-x}Co_xO_3$ (x = 0, 0.05, 0.10, 0.15) between 1.5 and 4.5 V at 5 mA·g^{-1}.

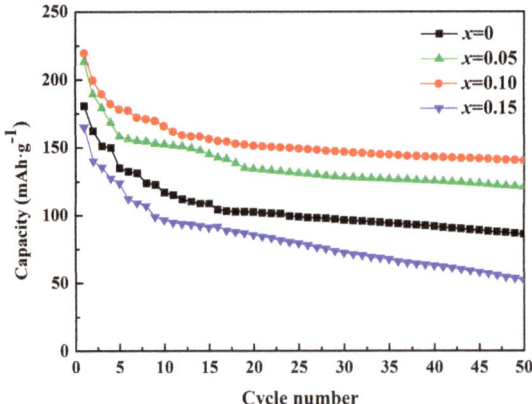

Figure 7. Cycling performances of the synthesized $Li_2Mo_{1-x}Co_xO_3$ (x = 0, 0.05, 0.10, 0.15) between 1.5 and 4.5 V at 5 mA·g^{-1}.

The comparison of rate capabilities between the pristine and Co-doped Li_2MoO_3 with current density from 5 mA·g^{-1} to 20 mA·g^{-1} is evaluated in Figure 8. The $Li_2Mo_{0.90}Co_{0.10}O_3$ possesses a higher discharge capacity of 218 mAh·g^{-1} at 5 mA·g^{-1} and 137 mAh·g^{-1} at 20 mA·g^{-1} than pristine Li_2MoO_3 material (180 mAh·g^{-1} at 5 mA·g^{-1} and 71 mAh·g^{-1} at 20 mA·g^{-1}). The difference of discharge capacities between Li_2MoO_3 and $Li_2Mo_{0.90}Co_{0.10}O_3$ increases from 38 mAh·g^{-1} to 66 mAh·g^{-1} with a rise of current density from 5 mA·g^{-1} to 20 mA·g^{-1}. When the current density returns to 5 mA·g^{-1}, the discharge capacity of $Li_2Mo_{0.90}Co_{0.10}O_3$ could reach up 155 mAh·g^{-1}, while that of pristine Li_2MoO_3 only lefts 104 mAh·g^{-1}. The above results suggest that Co-doping significantly enhances the rate capability of Li_2MoO_3.

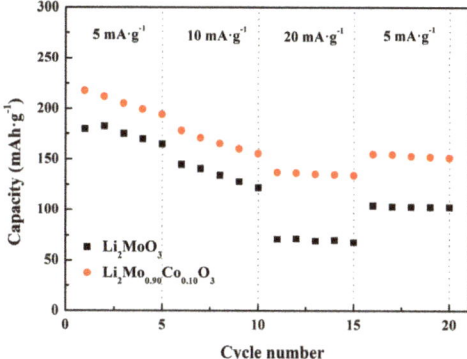

Figure 8. Rate performances of Li_2MoO_3 and $Li_2Mo_{0.90}Co_{0.10}O_3$ at different current density.

Figure 9 illustrates CV curves of Li_2MoO_3 and $Li_2Mo_{0.90}Co_{0.10}O_3$ between 1.5 and 4.5 V. The redox peaks of Li_2MoO_3 in CV curve are located at 2.965 V and 1.978 V, which are respectively related to the delithiation/lithiation processes corresponding to the oxidation/reduction of Mo^{4+}/Mo^{6+} couple [18]. The oxidation peak of the $Li_2Mo_{0.90}Co_{0.10}O_3$ at 2.863 V is lower than that of Li_2MoO_3 and the reduction peak of the $Li_2Mo_{0.90}Co_{0.10}O_3$ at 2.136 V is above that of the pristine sample. Therefore, $Li_2Mo_{0.90}Co_{0.10}O_3$ possesses a smaller difference of the redox peak potential (ΔE, 0.727 V) than Li_2MoO_3 (0.987 V). It is well known that difference of the redox peaks potential is highly correlated with electrode polarization. Hence, the conclusion can be drawn that Co-doping can reduce the polarization of Li_2MoO_3, which are coincident with the improvement in electrochemical performances.

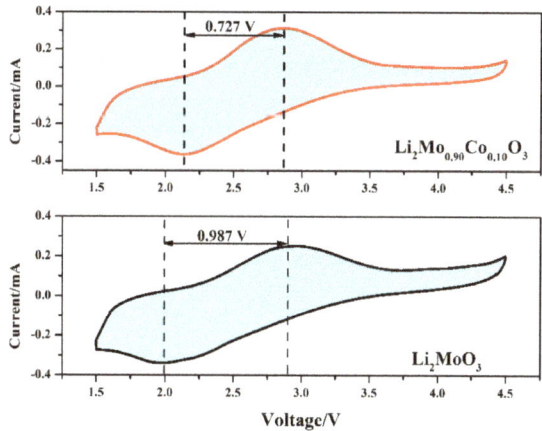

Figure 9. CV curves of Li_2MoO_3 and $Li_2Mo_{0.90}Co_{0.10}O_3$ between 1.5 and 4.5 V at the 5th cycle.

To further analyze the kinetic behaviors of the pristine and Co-doped samples, EIS measurements of Li_2MoO_3 and $Li_2Mo_{0.90}Co_{0.10}O_3$ were carried out and the results are presented in Figure 10. Both EIS plots display similar shapes (see Figure 10a), which are fitted through the equivalent circuit (see Figure 10c) and the fitting results are listed in Table 1. In the equivalent circuit, R_f is related to Li$^+$ diffusion in the SEI film, R_{ct} is corresponding to the charge transfer resistance at electrolyte-electrode interface and R_s is considered to be ohmic resistance. As can be seen from Table 1, the R_s and R_f of both samples change slightly. In contrast, the R_{ct} of Li_2MoO_3 is significantly reduced duo to the Co-doping. The $Li_2Mo_{0.90}Co_{0.10}O_3$ presents the R_{ct} of 105.70 Ω, which is far below the pristine Li_2MoO_3 (478.75 Ω). The R_{ct} is major part to the total electrode impedance, and its reduction reveals that Co-doping is very beneficial to enhance the kinetic behaviors of Li_2MoO_3. In addition, The Li^+ ion diffusion coefficients (D_{Li^+}) were estimated by the following formula:

$$D_{Li^+} = \frac{R^2T^2}{2A^2F^4C_{Li^+}^2\sigma^2} \quad (1)$$

R, T, A, F and C_{Li^+} are the gas constant, the absolute temperature, the area of the electrode surface, the Faraday's constant and the molar concentration of Li ions, respectively [7]. The σ corresponding to the Warburg factor could be calculated by the $Z'/\omega^{-0.5}$ (see Figure 10b) and the following formula:

$$Z' = R_f + R_{ct} + \sigma\omega^{-0.5} \quad (2)$$

The improving trend in the values of D_{Li^+} is very similar to the reducing trend in the values of R_{ct}. The pristine and Co-doped Li_2MoO_3 deliver the Li^+ ion diffusion coefficients of 3.89×10^{-17} and 1.94×10^{-16} cm$^2 \cdot$s^{-1}, respectively. As we can see, the Li^+ ion diffusion coefficients of Li_2MoO_3 have an obvious growth due to the Co-doping. These results clearly indicate that Co-doping significantly improves the kinetics behavior of Li^+ and reduces the impedance of Li_2MoO_3, which is responsible for the better rate capability and cycling stability of $Li_2Mo_{0.90}Co_{0.10}O_3$.

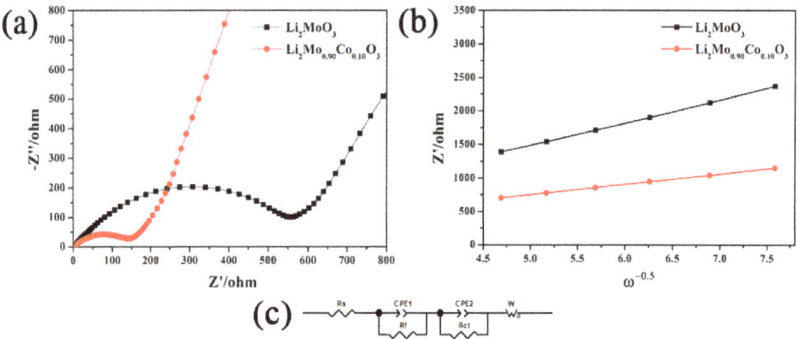

Figure 10. (a) EIS plots of Li_2MoO_3 and $Li_2Mo_{0.90}Co_{0.10}O_3$ after 20 cycles between 1.5 and 4.5 V at 5 mA·g^{-1}, (b) Z' vs. $\omega^{-0.5}$ at low frequency of EIS plots and (c) equivalent circuit model.

Table 1. Fitting results of EIS plots.

Samples	R_s/Ω	R_f/Ω	R_{ct}/Ω	D_{Li^+} (cm$^2 \cdot$s^{-1})
Li_2MoO_3	6.21	32.73	478.75	3.89×10^{-17}
$Li_2Mo_{0.90}Co_{0.10}O_3$	5.89	25.79	105.70	1.94×10^{-16}

To investigate the structural transformation of the samples during charge-discharge process, XRD patterns of Li_2MoO_3 and $Li_2Mo_{0.90}Co_{0.10}O_3$ after 20 cycles are exhibited in Figure 11. There are notable differences in the structure of Li_2MoO_3 before and after cycling. For pristine Li_2MoO_3 material, the strongest diffraction peak transfers from (003) to (104) and ratio of (003)/(104) drops to 0.65 after cycling, which could be indexed into the Li-insufficient structure [17]. This Li-insufficient structure contributes to the partially reversible migration of the Mo ions and the partial recovery of the Mo_3O_{13} clusters during charge-discharge processes, leading to irreversible capacity loss and poor cycling stability. In contrast, ratio of (003)/(104) of $Li_2Mo_{0.90}Co_{0.10}O_3$ after cycling can maintain at 0.83, which is much better than the pristine Li_2MoO_3 after cycling. It indicates that Co-doping can effectively enhance the structural stability during charge-discharge process.

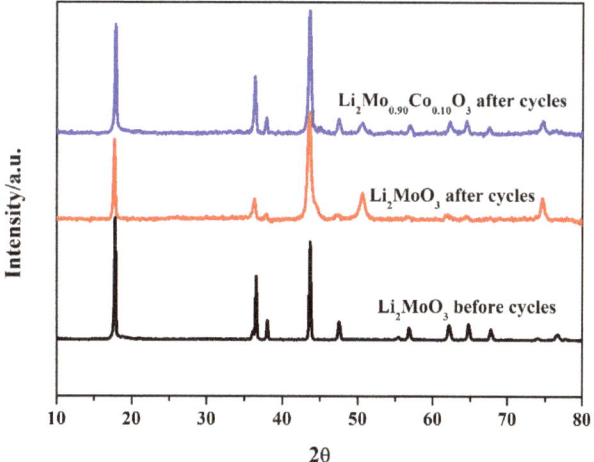

Figure 11. XRD patterns of Li_2MoO_3 and $Li_2Mo_{0.90}Co_{0.10}O_3$ after 20 cycles.

4. Conclusions

Co-doped Li_2MoO_3 was successfully synthesized via a solid phase method. The influences of Co-doping on the structural and electrochemical characteristics of Li_2MoO_3 are analyzed. The results show that the addition of Co affects the morphology and decreases the particle size of cathode materials. Electrochemical measurements confirm that Co-doping can effectively improve the electrochemical performances of Li_2MoO_3 materials. The $Li_2Mo_{0.90}Co_{0.10}O_3$ presents an initial discharge capacity of 220 mAh·g^{-1} with the capacity retention of 63.6% after 50 cycles at 5 mA·g^{-1}, which is much better than the pristine samples (181 mAh·g^{-1}, 47.5%). Additionally, the rate capability of Li_2MoO_3 is also enhanced by Co-doping. It is found that the $Li_2Mo_{0.90}Co_{0.10}O_3$ delivers a much lower R_{ct} and a higher Li^+ ion diffusion coefficients than pristine Li_2MoO_3. What is more, the irreversible structural transformation is also suppressed by Co-doping. The enhanced electrochemical performances could be attributed to the improvement in structural stability and reduction in impedance due to the Co-doping. Our work reveals that doping modification will be a promising method for improving the electrochemical performances of Li_2MoO_3 material, and thus benefit for its application.

Author Contributions: Conceptualization, Z.Y. and H.L.; methodology, J.H. and W.L.; formal analysis, J.H. and W.L.; writing-original draft preparation, J.H.; supervision, writing-review and editing, Z.Y.

Funding: This research was funded by Natural Science Foundation of China, grant number 51372191 and National Basic Research Program of China, grant number 2015CB656401.

Conflicts of Interest: The authors declare no conflict of interest.

References

1. Wang, Y.; Gu, H.T.; Song, J.H.; Feng, Z.; Zhou, X.; Zhou, Y.N.; Wang, K.; Xie, J. Suppressing Mn reduction of Li-rich Mn-based cathodes by F-doping for advanced lithium-ion batteries. *J. Phys. Chem. C* **2018**, *122*, 27836–27842. [CrossRef]
2. Ma, J.; Zhou, Y.N.; Gao, Y.; Kong, Q.; Wang, Z.; Yang, X.Q.; Chen, L. Molybdenum substitution for improving the charge compensation and activity of Li_2MnO_3. *Chem. Eur. J.* **2014**, *20*, 8723–8730. [CrossRef] [PubMed]
3. Zhao, W.; Xiong, L.; Xu, Y.; Xiao, X.; Wang, J.; Ren, Z. Magnesium substitution to improve the electrochemical performance of layered Li_2MnO_3 positive-electrode material. *J. Power Sources* **2016**, *330*, 37–44. [CrossRef]
4. Bareño, J.; Balasubramanian, M.; Kang, S.H.; Wen, J.G.; Lei, C.H.; Pol, S.V.; Petrov, I.; Abraham, D.P. Long-range and local structure in the layered oxide $Li_{1.2}Co_{0.4}Mn_{0.4}O_2$. *Chem. Mater.* **2011**, *23*, 2039–2050. [CrossRef]

5. Zhang, Q.; Peng, T.; Zhan, D.; Hu, X. Synthesis and electrochemical property of xLi$_2$MnO$_3$·(1 − x)LiMnO$_2$ composite cathode materials derived from partially reduced Li$_2$MnO$_3$. *J. Power Sources* **2014**, *250*, 40–49. [CrossRef]
6. Lanz, P.; Sommer, H.; Schulz, D.M.; Novák, P. Oxygen Release from high-energy xLi$_2$MnO$_3$·(1 − x)LiMO$_2$ (M = Mn, Ni, Co): Electrochemical, differential electrochemical mass spectrometric, in situ pressure, and in situ temperature characterization. *Electrochim. Acta* **2013**, *93*, 114–119. [CrossRef]
7. Kim, S.J.; Kim, M.C.; Kwak, D.H.; Kim, D.M.; Lee, G.H.; Choe, H.S.; Park, K.W. Highly stable TiO$_2$ coated Li$_2$MnO$_3$ cathode materials for lithium-ion batteries. *J. Power Sources* **2016**, *304*, 119–127. [CrossRef]
8. Wang, F.; Xiao, S.; Li, M.; Wang, X.; Zhu, Y.; Wu, Y.; Shirakawa, A.; Pe, J. A nanocomposite of Li$_2$MnO$_3$ coated by FePO$_4$ as cathode material for lithium ion batteries. *J. Power Sources* **2015**, *287*, 416–421. [CrossRef]
9. Xiao, R.; Li, H.; Chen, L. Density functional investigation on Li$_2$MnO$_3$. *Chem. Mater.* **2012**, *24*, 4242–4251. [CrossRef]
10. Zheng, J.; Gu, M.; Xiao, J.; Zuo, P.; Wang, C.; Zhang, J.G. Corrosion/fragmentation of layered composite cathode and related capacity/voltage fading during cycling process. *Nano Lett.* **2013**, *13*, 3824–3830. [CrossRef] [PubMed]
11. Bettge, M.; Li, Y.; Gallagher, K.; Zhu, Y.; Wu, Q.; Lu, W.; Bloom, I.; Abraham, D.P. Voltage fade of layered oxides: Its measurement and impact on energy density. *J. Electrochem. Soc.* **2013**, *160*, A2046–A2055. [CrossRef]
12. Pearce, P.E.; Perez, A.J.; Rousse, G.; Saubanère, M.; Batuk, D.; Foix, D.; McCalla, E.; Abakumov, A.M.; Tendeloo, G.V.; Doublet, M.L.; et al. Evidence for anionic redox activity in a tridimensional-ordered Li-rich positive electrode β-Li$_2$IrO$_3$. *Nat. Mater.* **2017**, *16*, 580–586. [CrossRef]
13. Miura, Y.; Yasui, Y.; Sato, M.; Igawa, N.; Kakurai, K. New-Ttype phase transition of Li$_2$RuO$_3$ with honeycomb structure. *J. Phys. Soc. Jpn.* **2007**, *76*, 033705. [CrossRef]
14. Arunkumar, P.; Jeong, W.J.; Won, S.; Im, W.B. Improved electrochemical reversibility of over-lithiated layered Li$_2$RuO$_3$ cathodes: Understanding aliovalent Co^{3+} substitution with excess lithium. *J. Power Sources* **2016**, *324*, 428–438. [CrossRef]
15. Takahashi, Y.; Kijima, N.; Hayakawa, H.; Awaka, J.; Akimoto, J. Single-crystal synthesis and structure refinement of Li$_2$MoO$_3$. *J. Phys. Chem. Solids* **2008**, *69*, 1518–1520. [CrossRef]
16. Ma, J.; Gao, Y.R.; Wang, Z.X.; Chen, L. Structural and electrochemical stability of Li-rich layer structured Li$_2$MoO$_3$ in air. *J. Power Sources* **2014**, *258*, 314–320. [CrossRef]
17. Ma, J.; Zhou, Y.N.; Gao, Y.R.; Yu, X.; Kong, Q.; Gu, L.; Wang, Z.; Yang, X.; Chen, L. Feasibility of using Li$_2$MoO$_3$ in constructing Li-rich high energy density cathode materials. *Chem. Mater.* **2014**, *26*, 3256–3262. [CrossRef]
18. Self, E.C.; Zou, L.; Zhang, M.J.; Opfer, R.; Ruther, R.E.; Veith, G.M.; Song, B.; Wang, C.; Wang, F.; Huq, A.; et al. Synthesis and electrochemical and structural investigations of oxidatively stable Li$_2$MoO$_3$ and xLi$_2$MoO$_3$·(1 − x)LiMO$_2$ composite cathodes. *Chem. Mater.* **2018**, *30*, 5061–5068. [CrossRef]
19. Lee, J.; Urban, A.; Li, X.; Su, D.; Hautier, G.; Ceder, G. Unlocking the potential of cation-disordered oxides for rechargeable lithium batteries. *Science* **2014**, *343*, 519–522. [CrossRef]
20. Kumakura, S.; Shirao, Y.; Kubota, K.; Komaba, S. Preparation and electrochemical properties of Li$_2$MoO$_3$/C composites for rechargeable Li-ion batteries. *Phys. Chem. Chem. Phys.* **2016**, *18*, 28556–28563. [CrossRef] [PubMed]
21. Yu, Z.; Yu, T.; Li, W.; Hao, J.; Liu, H.; Sun, N.; Lu, M.; Ma, J. Improved electrochemical performances of carbon-coated Li$_2$MoO$_3$ cathode materials for Li-ion batteries. *Int. J. Electrochem. Sci.* **2018**, *13*, 4504–4511. [CrossRef]
22. Yuan, B.; Liao, S.X.; Xin, Y.; Zhong, Y.; Shi, X.; Li, L.; Guo, X. Cobalt-doped lithium-rich cathode with superior electrochemical performance for lithium-ion batteries. *RSC Adv.* **2015**, *5*, 2947–2951. [CrossRef]
23. Oz, E.; Demirel, S.; Altin, S. Fabrication and electrochemical properties of LiCo$_{1-x}$Ru$_x$O$_2$ cathode materials for Li-ion battery. *J. Alloys Compd.* **2016**, *671*, 24–33. [CrossRef]
24. Tang, Z.; Wang, Z.; Li, X.; Peng, W. Preparation and electrochemical properties of Co-doped and none-doped Li[Li$_x$Mn$_{0.65(1-x)}$Ni$_{0.35(1-x)}$]O$_2$ cathode materials for lithium battery batteries. *J. Power Sources* **2012**, *204*, 187–192. [CrossRef]

25. Song, J.; Shao, G.; Shi, M.; Ma, Z.; Song, W.; Wang, C.; Liu, S. The effect of doping Co on the electrochemical properties of LiFePO$_4$/C nanoplates synthesized by solvothermal route. *Solid State Ion.* **2013**, *253*, 39–46. [CrossRef]
26. Kim, Y. First principles investigation of the structure and stability of LiNiO$_2$ doped with Co and Mn. *J. Mater. Sci.* **2012**, *47*, 7558–7563. [CrossRef]
27. Becker, D.; Cherkashinin, G.; Hausbrand, R.; Jaegermann, W. Adsorption of diethyl carbonate on LiCoO$_2$ thin films: Formation of the electrochemical interface. *J. Phys. Chem. C* **2014**, *118*, 962–967. [CrossRef]
28. Guan, J.; Li, Y.; Guo, Y.; Su, R.; Gao, G.; Song, H.; Yuan, H.; Liang, B.; Guo, Z. Mechanochemical process enhanced cobalt and lithium recycling from wasted lithium-ion batteries. *ACS Sustain. Chem. Eng.* **2017**, *5*, 1026–1032. [CrossRef]
29. Li, X.; Qu, M.; Yu, Z. Structural and electrochemical performances of Li$_4$Ti$_{5-x}$Zr$_x$O$_{12}$ as anode material for lithium-ion batteries. *J. Alloys Compd.* **2009**, *487*, L12–L17. [CrossRef]
30. Zhao, R.; Hung, I.M.; Li, Y.T.; Chen, H.; Lin, C.P. Synthesis and properties of Co-doped LiFePO$_4$ as cathode material via a hydrothermal route for lithium-ion batteries. *J. Alloys Compd.* **2012**, *513*, 282–288. [CrossRef]
31. Yabuuchi, N.; Tahara, Y.; Komaba, S.; Kitada, S.; Kajiya, Y. Synthesis and electrochemical properties of Li$_4$MoO$_5$-NiO binary system as positive electrode materials for rechargeable lithium batteries. *Chem. Mater.* **2016**, *28*, 416–419. [CrossRef]
32. Kim, D.Y.; Ahn, H.J.; Kim, J.S.; Kim, I.P.; Kweon, J.H.; Nam, T.H.; Kim, K.W.; Ahn, J.H.; Hong, S.H. The Electrochemical properties of nano-sized cobalt powder as an anode material for lithium batteries. *Electron. Mater. Lett.* **2009**, *5*, 183–186. [CrossRef]

© 2019 by the authors. Licensee MDPI, Basel, Switzerland. This article is an open access article distributed under the terms and conditions of the Creative Commons Attribution (CC BY) license (http://creativecommons.org/licenses/by/4.0/).

Article

Preparation of LiFePO$_4$/C Cathode Materials via a Green Synthesis Route for Lithium-Ion Battery Applications

Rongyue Liu [1,2,*,†], Jianjun Chen [1,*], Zhiwen Li [1,3,†], Qing Ding [2], Xiaoshuai An [3], Yi Pan [2], Zhu Zheng [2], Minwei Yang [2] and Dongju Fu [1]

1. Research Institute of Tsinghua University in Shenzhen, High-Tech Industry Park, Nanshan District, Shenzhen 518057, China; zwlihit@163.com (Z.L.); youyou.orange23@163.com (D.F.)
2. Shenzhen Institute of THz Technology and Innovation, Xixiang, Bao'an District, Shenzhen 518102, China; dingqing@huaxunchina.cn (Q.D.); panyi@huaxunchina.cn (Y.P.); zzealot99@gmail.com (Z.Z.); yangpound@163.com (M.Y.)
3. Shenzhen Graduate School, Harbin Institute of Technology, Shenzhen University Town, Xili, Shenzhen 518055, China; anxiaoshuai@126.com
* Correspondence: liuryu@163.com (R.L.); chenjj08@126.com (J.C.)
† The authors equally contributed to this work.

Received: 12 October 2018; Accepted: 9 November 2018; Published: 12 November 2018

Abstract: In this work, LiFePO$_4$/C composite were synthesized via a green route by using Iron (III) oxide (Fe$_2$O$_3$) nanoparticles, Lithium carbonate (Li$_2$CO$_3$), glucose powder and phosphoric acid (H$_3$PO$_4$) solution as raw materials. The reaction principles for the synthesis of LiFePO$_4$/C composite were analyzed, suggesting that almost no wastewater and air polluted gases are discharged into the environment. The morphological, structural and compositional properties of the LiFePO$_4$/C composite were characterized by X-ray diffraction (XRD), scanning electron microscope (SEM), transmission electron microscopy (TEM), Raman and X-ray photoelectron spectroscopy (XPS) spectra coupled with thermogravimetry/Differential scanning calorimetry (TG/DSC) thermal analysis in detail. Lithium-ion batteries using such LiFePO$_4$/C composite as cathode materials, where the loading level is 2.2 mg/cm^2, exhibited excellent electrochemical performances, with a discharge capability of 161 mA h/g at 0.1 C, 119 mA h/g at 10 C and 93 mA h/g at 20 C, and a cycling stability with 98.0% capacity retention at 1 C after 100 cycles and 95.1% at 5 C after 200 cycles. These results provide a valuable approach to reduce the manufacturing costs of LiFePO$_4$/C cathode materials due to the reduced process for the polluted exhaust purification and wastewater treatment.

Keywords: LiFePO$_4$/C composite; cathode material; green synthesis route; lithium-ion batteries

---

Highlights:

✔ LiFePO$_4$/C is synthesized by using a green route where almost no wastewater and air polluted gases are discharged into the environment.
✔ The reaction principles for the synthesis of LiFePO$_4$/C are analyzed.
✔ LiFePO$_4$/C exhibits uniform nano-structure and carbon layer.
✔ LiFePO$_4$/C shows excellent rate capability and cycling capability.

1. Introduction

Olivine-type LiFePO$_4$ is considered as one of the most promising cathode materials for Li ions batteries owing to its high operating voltage (~3.4 V vs. Li/Li$^+$), high theoretical capacity (~170 mA h/g), low cost and no environmental pollution [1–7]. However, bare LiFePO$_4$ materials suffer from many disadvantages, such as low conductivity and sluggish diffusion rate of Li$^+$ ions coupled with low tap density [6,7]. Recently, many efforts have been made to improve its conductivity and accelerate the diffusion rate of Li$^+$, including coating the conducting materials on the surface of LiFePO$_4$ materials [8–11], reducing particle size [12], doping transition metals ions [13], etc. Ultimately, high-quality LiFePO$_4$ materials have been successfully developed and commercialized in energy storage and electric vehicles (EVs).

However, there still exist some challenging problems for the commercialization of LiFePO$_4$ materials in the next generation of lithium-ion batteries. Firstly, complex fabrication procedures such as ingredients, pulping, coating, tableting, winding and assembly welding, further need to be simplified and optimized [7,14]. Secondly, understanding the kinetic behavior of LiFePO$_4$ material for the lithium-ion batteries is of fundamental importance [2,6], including the conductive pathway with conducting materials coated on its surface, the Li$^+$ ions diffusion dynamics with the transition metal atoms doping, the Li$^+$ ions diffusion pathway during the insertion/extraction process, etc. Thirdly, reducing the manufacturing costs of LiFePO$_4$ materials and preventing environmental pollution are quite important. Currently, it is noted that the synthesis LiFePO$_4$ materials always uses solid state reaction method [15,16], liquid phase method [17], sol-gel method [18], hydrothermal method [19,20] and spray pyrolysis method [21,22]. Almost all these methods can produce wastewater containing excessive anions impurities such as SO_4^{2-}, Cl^- and NO_3^-, and contaminated gas (N_xO_y, CO, and NH_3), which need additional apparatus to deal with them and increase the manufacturing cost. Therefore, seeking approaches to further reduce the manufacturing cost of LiFePO$_4$ materials synthesis and preventing environmental pollution are still highly pursued by materials scientists.

Here, we developed a green route to synthesize the LiFePO$_4$/C composite by using Iron (III) oxide (Fe$_2$O$_3$) nanoparticles, Lithium carbonate (Li$_2$CO$_3$), glucose powder and phosphoric acid (H$_3$PO$_4$) solution as raw materials. We first synthesized the FePO$_4$·2H$_2$O precursor by the reaction of Fe$_2$O$_3$ nanoparticles with H$_3$PO$_4$ solution. The wastewater was water and excessive H$_3$PO$_4$ solution which could be recycled next time. Second, we synthesized the LiFePO$_4$/C composite by annealing the mixtures composed of FePO$_4$·2H$_2$O precursor, Li$_2$CO$_3$ and glucose powder at a high-temperature process, where only CO$_2$ gas and water vapor were discharged. Therefore, all the reaction processes were environmentally friendly. The morphological, structural, compositional properties of the synthesized LiFePO$_4$/C composite were characterized. Lithium-ion batteries using such composite as cathode active materials were fabricated, and the corresponding electrochemical performance were discussed.

2. Experimental Section

2.1. Preparation of LiFePO$_4$/C Composite

Iron (III) oxide (Fe$_2$O$_3$) powder (~800 nm (ACS, 99.99%)), phosphoric acid (H$_3$PO$_4$) solution (85%), Lithium carbonate (Li$_2$CO$_3$) and glucose powder were purchased from Sigma Aldrich (Shanghai, China) and used without further purification unless stated otherwise. The FePO$_4$·2H$_2$O precursor was prepared by the chemical reaction of Fe$_2$O$_3$ powder with H$_3$PO$_4$ solution where the molar ratio of the Fe/P was 1:1.05. In a typical procedure, 16 g of Fe$_2$O$_3$ powder and 14.4 mL H$_3$PO$_4$ solution were added into 20 mL deionized water in a flask followed by ultrasonic dispersion for 30 min. Then, the mixed slurries were transferred into a ball mill tank and ball-milled for additional 9 h. After that, the mixed slurries were filtered and allowed to heat up to 85 °C for 5 h forming a suspension, followed by cooling down to room temperature. The white precipitate (FePO$_4$·2H$_2$O precursor) was collected and separated by centrifugation and washed with water for several times, and then dried in a blast drying box for 24 h. The LiFePO$_4$/C composite were prepared by using stoichiometric amounts of FePO$_4$·2H$_2$O precursor, Li$_2$CO$_3$ and glucose powder (60.0 g glucose/1 mol FePO$_4$·2H$_2$O precursor) as the starting materials, followed by high temperature sintering. First, stoichiometric amounts of FePO$_4$·2H$_2$O precursor and Li$_2$CO$_3$ were added into water-dissolved glucose solution in a flask followed by ultrasonic dispersion for 30 min. Then, the mixed slurries were dried in a blast drying box for 24 h. After that, the mixture was sintered at 650 °C in a tube furnace for 10 h under argon flow to obtain LiFePO$_4$/carbon composite.

2.2. Characterization of LiFePO$_4$/C Composite

The crystallinity was estimated by using X-ray diffraction (XRD, D/Max-IIIC, Rigaku Co., Tokyo, Japan) equipped with a Cu-Kα source of wavelength λ = 1.54060 Å and operated at 40 kV and 20 mA. The top-view SEM images were taken on a Hitachi S-4800 (Hitachi Limited, Tokyo, Japan), and the attached energy dispersive spectrometer (EDS (Hitachi Limited, Tokyo, Japan) analyzer was used to analyze the composition distribution of carbon. The transmission electron microscopy (TEM) images were acquired on a FEI Talos F200X (FEI, Hillsboro, OR, USA) with an acceleration voltage of 200 kV. Thermo-Gravimetric coupled with Differential Scanning Calorimetry (TG-DSC, Netzsch Scientific Instruments Trading Co., Ltd., Shanghai, China) was used to measure the carbon content in the LiFePO$_4$/carbon composite. Raman spectra were tested at room temperature equipping with 514 nm laser excitations. X-ray Photoelectron Spectroscopy (XPS) measurement was performed on a SPECS HSA-3500 (SPECS, Berlin, Germany) to determine the valence state of each element of the samples.

2.3. Electrochemical Measurements

The electrochemical measurements were performed using a CR2032 coin-type cell assembled in an argon-filled glove-box (MIKROUNA, Guangzhou, China). For fabricating the working electrodes, a mixture of active materials (LiFePO$_4$/C composite), conductive carbon blacks (Super-P, Shenzhen, China), and polyvinylidene fluoride (PVDF) binder at a weight ratio of 80:10:10 was coated on aluminum foil and dried in vacuum at 120 °C for 12 h. The thickness and loading level were 48 μm and 2.2 mg/cm^2, respectively. The lithium pellets were used as the counter and reference electrode. The electrolyte consisted of a solution of 1 mol/L LiPF$_6$ in ethylene carbon (EC)/dimethyl carbonate (DMC) (1:1 w/w). A celguard 2300 microporous film was used as separator. The cells were galvanostatically charged and discharged between 2.5 V and 4.2 V versus Li/Li$^+$ on a battery cycler (LAND, CT2001A, Wuhan, China). Cyclic voltammogram (CV) measurements were carried out using a Multi Autolab electrochemical workstation (Metrohm, Guangzhou, China) at a scanning rate of 0.1–0.5 mV s^{-1}. Electrochemical impedance spectra (EIS) were also characterized by Autolab electrochemical workstation adjusting amplitude signal at 5 mV and frequency range of 0.01 Hz–100 kHz.

3. Results and Discussions

Figure 1 shows the schematic diagram of the preparation of LiFePO$_4$/C composite. Firstly, an excess of phosphoric acid solution reacted with Iron (III) oxide (Fe$_2$O$_3$) powder to form FePO$_4$·2H$_2$O precursors. Subsequently, these precursors were mixed with Lithium carbonate (Li$_2$CO$_3$) and glucose powder followed by high temperature sintering to form LiFePO$_4$/C composite. The reaction equations are shown below.

$$Fe_2O_3 + 2H_3PO_4 + H_2O \rightarrow FePO_4 \cdot 2H_2O. \tag{1}$$

$$2FePO_4 \cdot 2H_2O + Li_2CO_3 + C_6H_{12}O_6 \rightarrow 2LiFePO_4/C + \text{volatile matter} \tag{2}$$

Equation (1) shows the synthesis process of FePO$_4$·2H$_2$O precursors. The reaction mechanism is referred to the reported literature [23], where the Fe salts are used and the reaction equations are shown below:

$$Fe^{3+} + H_2O = Fe(OH)^{2+} + H^+ \tag{3}$$

$$Fe(OH)^{2+} + H_2O = Fe(OH)_2^+ + H^+ \tag{4}$$

$$Fe^{3+} + H_3PO_4 = FeH_2PO_4^{2+} + H^+ \tag{5}$$

$$FeH_2PO_4^{2+} + H_3PO_4 = Fe(H_2PO_4)_2^+ + H^+ \tag{6}$$

$$Fe(OH)^+ + Fe(H_2PO_4)_2^+ = 2FePO_4 + 2H_2O + 2H^+ \tag{7}$$

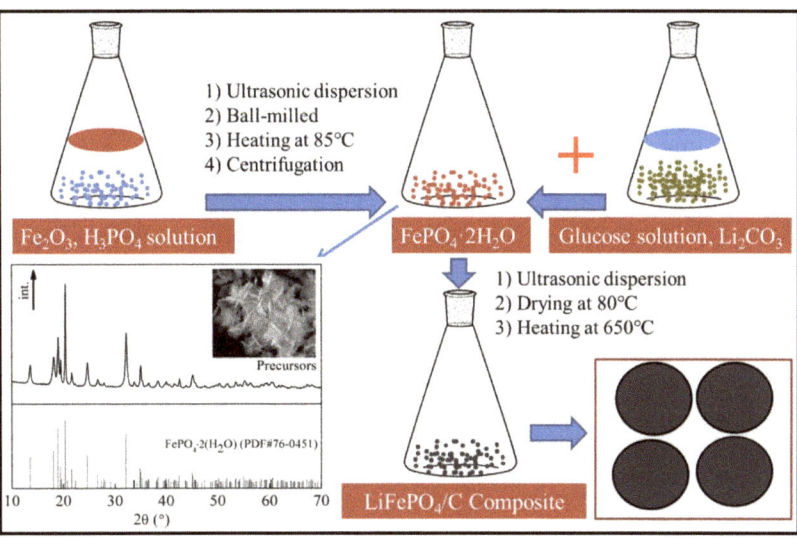

Figure 1. Schematic illustration for the preparation of the LiFePO$_4$/C composite.

In this work, we replaced the Fe salts with the Fe$_2$O$_3$ powder as raw materials to synthesize the FePO$_4$·2H$_2$O precursors because the wastewater from the above method contains impurity ions such as SO$_4^{2-}$, Cl$^-$ and NO$_3^-$ which are not environmentally friendly, although Fe salts are very cheap. Due to the complex reaction processes, we add Equations (3)–(7) to get Equation (1) where the Fe^{3+} ions are replaced by Fe$_2$O$_3$. In addition, it takes a little time to dissolve the Fe$_2$O$_3$ powder in the acid solution, leading to a lower reaction rate of our method in comparison with that of the method described above. In the environmental protection perspective, our method is favorable to the commercialization of future products. The reason is that there are no metal ion and anion impurities left in the wastewater solution. Although the phosphoric acid solution is excessive, it can be recovered and recycled. Equation (2) displays the synthesis process of LiFePO$_4$/C composite. The reaction equation is balanced according to the stoichiometric values of Li, Fe and P elements, where LiFePO$_4$/C is the final product while the volatile matter represents the volatile gases, such as CO$_2$ gas and H$_2$O vapor, even a small amount of CO gas, other C$_x$H$_y$O$_z$, etc. To determine whether the CO gas or C$_x$H$_y$O$_z$ is present in the exhaust during the synthesis of LiFePO$_4$/C composite, thermodynamic Gibbs free energies for the formation of CO and C$_x$H$_y$O$_z$ were calculated. Because the precursors were annealed in inert gas, the thermal decomposition products of the FePO$_4$·2H$_2$O are FePO$_4$ and H$_2$O vapor, while those of Li$_2$CO$_3$ are Li$_2$O and CO$_2$, and those of C$_6$H$_{12}$O$_6$ are C and H$_2$O vapor. Therefore, the formation of CO or C$_x$H$_y$O$_z$ can be derived from the reaction of C with H$_2$O vapor or CO$_2$ gas. By thermodynamic Gibbs free energies calculation, the temperature for the formation of CO gas must be more than 980.6 K. In our experiment, the sintering temperature for the precursors was 650 °C (923K), i.e. lower than 980.6 K. Thus, CO gas was not generated. C$_x$H$_y$O$_z$ was also not generated due to absence of CO and H$_2$. In other words, our work is a green route to synthesize the LiFePO$_4$/C cathode materials for lithium ion battery application.

Figure 2a shows the XRD pattern of LiFePO$_4$/C composite. As can be seen, all XRD peaks match well with the standard data JCPDS (Joint Committee on Powder Diffraction Standards) card

No. 81-1173, demonstrating the formation of LiFePO$_4$ with orthorhombic structure. The lattice parameters are a = 10.342 Å, b = 6.021 Å, and c = 4.699 Å, respectively. The main XRD peaks are strong and sharp, suggesting good crystallinity of LiFePO$_4$/C composite. The XRD peaks assigned to the carbon are not detected due to its amorphous state [22]. Moreover, its low content also plays an important role. Figure 2b shows the TG/DSC curves to estimate the carbon content in the LiFePO$_4$/C composite. As can be seen, the weight gain of 3.62% below 550 °C is assigned to the oxidation of LiFePO$_4$/carbon to the Li$_3$Fe$_2$(PO$_4$)$_3$ and Fe$_2$O$_3$ [24,25]. Above 550 °C, there is almost no weight change, indicating that the LiFePO$_4$/C composite are fully oxidized where the carbon is oxidated to the CO$_2$ gas. According to the total weight gain of 5.07% for pure LiFePO$_4$ in theory [26], the amount of carbon in the LiFePO$_4$/C composite is about 1.45%. Figure 2c shows the Raman characterization of LiFePO$_4$/C composite. The Raman spectrum exhibits two peaks at 1351 cm^{-1} and 1605 cm^{-1} corresponding to the D band (disordered carbon, sp3) and G band (graphite, sp2) for amorphous carbon, respectively [27–30]. The observed D band and G band indicate the existence of carbon in the LiFePO$_4$/C composite. A lower relative intensity ratio of D/G band corresponds to a higher order carbon arrangement. As can be seen, the relative intensity ratio of D/G is 0.66 and the G band shows a smaller full-width half-maximum compared to that of the D band, indicating high graphitization of C in the LiFePO$_4$/C composite. Although the Raman spectrum shows a sharp graphitic carbon peak, the carbon remains in the amorphous state in the LiFePO$_4$/C composite. Therefore, it is not detected by XRD characterization. Figure 2d shows EDS mapping to estimate the composition distribution of carbon element in the LiFePO$_4$/C composite. As can be seen, the carbon is uniformly distributed across the whole surface, which is beneficial to the conductivity properties of LiFePO$_4$ and improves electrochemical performance of Lithium-ion battery.

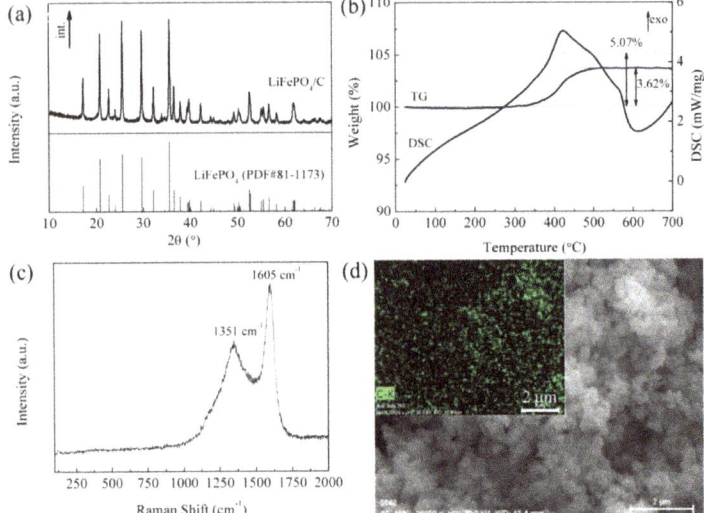

Figure 2. (**a**) XRD pattern of LiFePO$_4$/C composite; (**b**) TG-DSC curves of the LiFePO$_4$/C composite recorded from the room temperature to 700 °C at a heating rate of 10 °C min^{-1} in air; (**c**) Raman spectrum of LiFePO$_4$/C composite; and (**d**) EDS mapping of C in the LiFePO$_4$/C composite.

Figure 3a shows the SEM images of LiFePO$_4$/C composite. As can be seen, the LiFePO$_4$/C composite exhibit uniform particle size distribution ranging from 100 to 200 nm. The small grain sizes of LiFePO$_4$/C composite are attributed to the carbon coating on the surface of the LiFePO$_4$ nanoparticles that prevents their quick growth. This phenomenon can be explained by the space steric effect which increases the diffusion activation energy of the reactants and slows down the growth rate of grains [31]. Therefore, the carbon coating layer is quite important in controlling particle size. The small grain sizes are conducive to shortening the migration paths of lithium ions and electrons during the lithiation/delithiation process and as a result, improve the electrochemical performances of LiFePO$_4$/C composite efficiently [32]. Further characterization was carried out by TEM and the corresponding images of the LiFePO$_4$/C composite are shown in Figure 3b–d. The carbon layer on the LiFePO$_4$ nanoparticles surface is uniform, showing a thickness of about 2–3 nm, which demonstrates that the carbon exists in the LiFePO$_4$/C composite. This result is consistent with the previous TG-DSC analysis and Raman characterization. The effect of the carbon layer is beneficial to smoothing electron migration for the reverse reaction of Fe^{3+} to Fe^{2+}. In addition, the carbon layer can supply a better electronic contact between the LiFePO$_4$ nanoparticles, which ensures that the electrons are able to migrate quickly enough from all sides [32–34]. Meanwhile, the lattice fringes corresponding to the (011) crystal plane demonstrate the formation of olivine-type LiFePO$_4$.

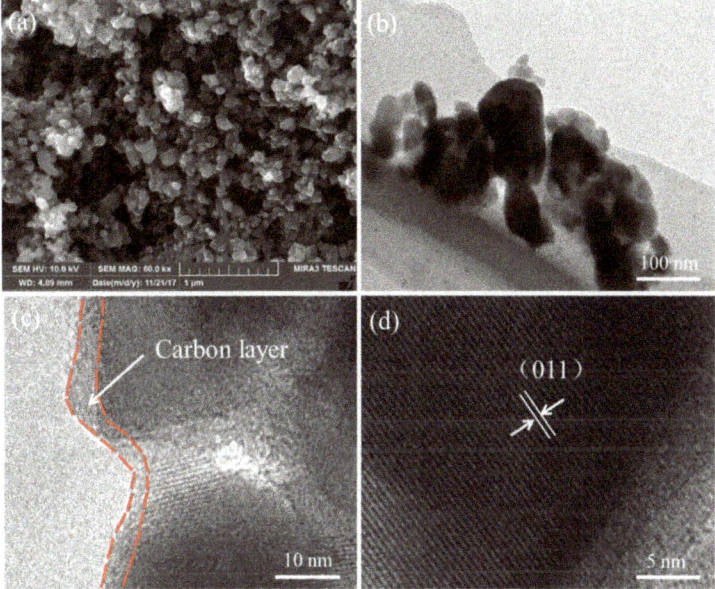

Figure 3. (**a**) SEM image of LiFePO$_4$/C composite; and (**b**–**d**) TEM images of LiFePO$_4$/C composite.

Figure 4 shows the high-resolution X-ray photoelectron spectroscopy (XPS) spectra of the Li 1s, Fe 3p, Fe 2p, P 2p, O 1s and C 1s core levels to determine the oxidation states of the elements in the LiFePO$_4$/C composite. The peak at 56.5 eV, corresponding to the lithium of the LiFePO$_4$/C composite, cannot be seen due to the superposed iron peak of Fe 3p [35,36]. The peak intensity of Fe 3p is higher than Li 1s because the Fe 3p has greater relative atomic sensitivity than that of Li 1 s [37,38]. The Fe 2p shows two peaks at 710.1 (2p3/2) and 724.1 eV (2p1/2) with a splitting energy of 14.0 eV, which is close to the standard splitting energy of 19.9 eV, demonstrating the oxidation state of Fe^{2+} [36,38]. Moreover, two small peaks at high binding energy of 713.9 and 728.5 eV are the characters of transition metal ions with partially filled-d orbits, which are assigned to the multiple splitting of the energy levels of Fe ion [37,38]. The peaks representing the other valence states of Fe ions cannot be seen, revealing that only Fe^{2+} ions exist in the LiFePO$_4$/C composite. The P 2p shows a peak at 132.9 eV, revealing that the valence state of P is 5+ [38]. The O 1s shows a peak at 531.0 eV, confirming that the valence state of O in the LiFePO$_4$/C composite is divalent. The two shoulder peaks at 531.9 and 533.0 eV are attributed to the C–O and C=O bands arising from functional groups absorbed on the sample surface [39]. The C 1s shows peaks at 284.0 and 284.4 eV, which correspond to the short-order sp2-coordinated and sp3-coordinated carbon atoms [38]. The additional peak at 288.2 eV is the C=O band arising from functional groups absorbed on the sample surface. These results confirm that the LiFePO$_4$/C composite was synthesized.

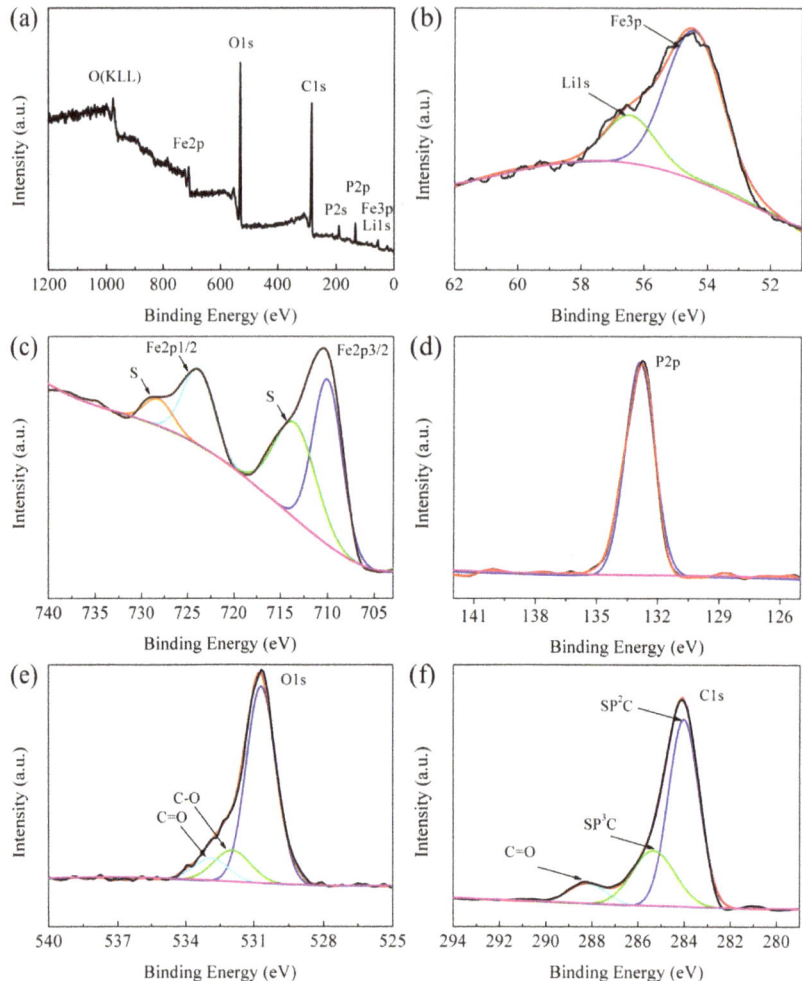

Figure 4. XPS survey of LiFePO$_4$/C composite (**a**); high resolution XPS spectrum of: Li 1s (**b**); Fe 2p (**c**); P 2p (**d**); O 1s (**e**); and C 1s (**f**) for LiFePO$_4$/C composite.

Figure 5a shows the cyclic voltammetry curves of lithium ion batteries using the LiFePO$_4$/C composite as the cathode active materials. N peak appears at 2.63 V (characteristic of Fe^{3+} in Fe$_2$O$_3$), indicating that all the iron atoms in the LiFePO$_4$/C composite are Fe^{2+} [40]. The two peaks around at 3.34 and 3.53 V (vs. Li$^+$/Li) are attributed to the Fe^{2+}/Fe^{3+} redox reaction, which corresponds to lithium extraction and insertion in LiFePO$_4$ crystal structure [41]. Furthermore, the two peaks show a narrow potential separation of 0.19 V and exhibit good symmetric and poignant shape, which imply a good electrochemical performance for lithium ion batteries. Figure 4b further shows the evolution of the cyclic voltammetry curves of LiFePO$_4$/C composite in the scanning rate ranging from 0.1 to 0.5 mV·s^{-1}. The peak position shifts and the potential separation between two peaks broadens gradually as the scan rate increases. Previous literature has reported that the diffusion coefficient of lithium ions (D_{Li}) can be determined from a linear relationship between peak currents (i_p) and the square root of the scan rate ($v^{1/2}$) based on the Randles–Sevcik equation [41–43]:

$$I_p = 2.69 \times 10^5 n^{3/2} ACD^{1/2} v^{1/2} \tag{8}$$

where I_p (A) is the current maximum, n is the number of electrons transfer per mole (n = 1), F (C/mol) is the Faraday constant, A (cm^2) is the electrode area (1.77 cm^2), C (mol/cm^3) is the lithium concentration in the LiFePO$_4$/C composite, v (V/s) is the scanning rate, D_{Li} (cm^2/s) is the lithium diffusion coefficient, R (J/K·mol) is the gas constant, and T (K) is the temperature. Figure 4c shows the linear relationship between peak currents (I_p) and the square root of the scan rate ($v^{1/2}$). The diffusion coefficient D_{Li} are calculated to be 4.35×10^{-13} and 2.57×10^{-13} cm^2/s for the charge and discharge processes, respectively, which are comparable to the previous reported literature [43–45]. This confirms that Li ions show excellent transmission performance, suggesting excellent electrochemical performance of our Li-ion batteries. Figure 4d,e shows the charge/discharge curves of lithium ion batteries at current rate from 0.1 C to 20 C. Apparently, at a low current rate of 0.1 C, the batteries deliver a discharge capacity of 161 mAh·g^{-1}, corresponding to 95% of the theoretical capacity (170 mAh·g^{-1}) of LiFePO$_4$. With the current rate increasing, the discharge capacity continually decreases, which is attributed to the low electronic conductivity and ion diffusion coefficient coupled with low tap density [32,38]. Despite this, the discharge capacity of our lithium ion batteries can reach 119 and 93 mAh·g^{-1} at high current rate of 10 C and 20 C. In addition, our batteries retain an approximate discharge capacity of 161 mAh·g^{-1} at the current rate of 0.1 C after the batteries are tested at the current rate of 20 C. This indicates that our batteries are highly structural stability, which can be suitable for the large current discharge. Figure 4f displays the cyclic performances and the coulombic efficiency of the lithium ion batteries. It is found that the batteries show a discharge capacity of 142 mAh·g^{-1} with a capacity retention of 98% after 100 cycles at 1 C. When the rate reaches at 5 C, the batteries even show discharge capacity of 125 mAh·g^{-1} with a capacity retention of 95.1% after 200 cycles. The coulombic efficiency with a value of 99% almost remains constant. These results demonstrate the high cycling stability of our batteries.

The electrochemical impedance spectra (EIS) technology is one of the most powerful tools to study electrochemical reactions, such as the processes occurring at the interface between electrodes and electrolyte, and the Li$^+$ intercalation/de-intercalation in the interior of cathode/anode materials [46,47]. Figure 6a shows the EIS curve of lithium ion batteries using the LiFePO$_4$/C composite as the cathode active materials after 10 cycles at rate of 1 C. Clearly, the EIS curve consists of a semicircle in the high-frequency region followed by a straight line in the low-frequency region. The former is related to the charge-transfer process at the electrode/electrolyte interfaces, while the latter represents the Warburg impedance associated with the Li$^+$ diffusion in the LiFePO$_4$ crystal lattice [48,49]. The radius of the semicircle in the EIS curve for the LiFePO$_4$/C composite is 60.2 Ω. As a comparison, the EIS curve of the commercial LiFePO$_4$/C materials is also plotted in Figure 6a. All the procedures for the fabrication of lithium ion batteries are completed under identical conditions. In addition, the loading level of commercial LiFePO$_4$/C composite as active materials is also 2.2 g/cm^2. The commercial LiFePO$_4$/C materials with the carbon content of about 1.44% are purchased from the Optimumnano Energy Co., Ltd. (Shenzhen, China). The grain size of the LiFePO$_4$/C is 200–300 nm, as shown in Figure 6b. As can be seen, the radius of the semicircle in the EIS curve is 124.2 Ω. This indicates that our LiFePO$_4$/C composite shows better electrical properties than that of the commercial LiFePO$_4$/C materials. One of the possible reasons is that our LiFePO$_4$/C composite (100–200 nm) exhibits relatively smaller grain sizes and higher specific surface area (Figure 6b) in comparison with that of the commercial LiFePO$_4$/C materials. This is because the small grain sizes are conducive to shortenig the migration paths of lithium ions and electrons during the lithiation/delithiation process [38]. In addition, the carbon content is very similar between our LiFePO$_4$/C composite and the commercial

LiFePO$_4$/C materials. The diffusion coefficient of Li$^+$ (D) can also be calculated form the EIS curve by using the following equation [49,50]:

$$D = R^2T^2/2A^2n^4F^4C^2\sigma^2 \qquad (9)$$

where R is gas constant (8.314 J·mol^{-1}·k^{-1}), T is the absolute temperature (298.15 K), A is the area of the tested electrode surface (cm^2), n is the number of electrons involved in the redox process (n = 1 in this work), C is the molar concentration of Li$^+$ in the tested electrode, F is the Faraday constant, and σ is the Warburg impedance coefficient [46,47]. By linear fitting the relation plot between Z_{Re} and $\omega^{-1/2}$ (the reciprocal square root of the angular frequency ω) (as shown in Figure 6b) to estimate the Warburg impedance coefficient σ, the diffusion coefficient of Li$^+$ (D) could be obtained from the above equation. By calculation, the diffusion coefficient of Li$^+$ (D) for our LiFePO$_4$/C composite is 3.17 × 10^{-13} cm^2/s. This result is consistent with the previous calculation using the Randles–Sevcik equation. The D value for the commercial LiFePO$_4$/C materials is also calculated to be 2.34 × 10^{-13} cm^2/s. For a comparison, our LiFePO$_4$/C composite shows a relatively higher D value, which is assigned to the smaller grain sizes that are conducive to shortening the migration paths of lithium ions [38].

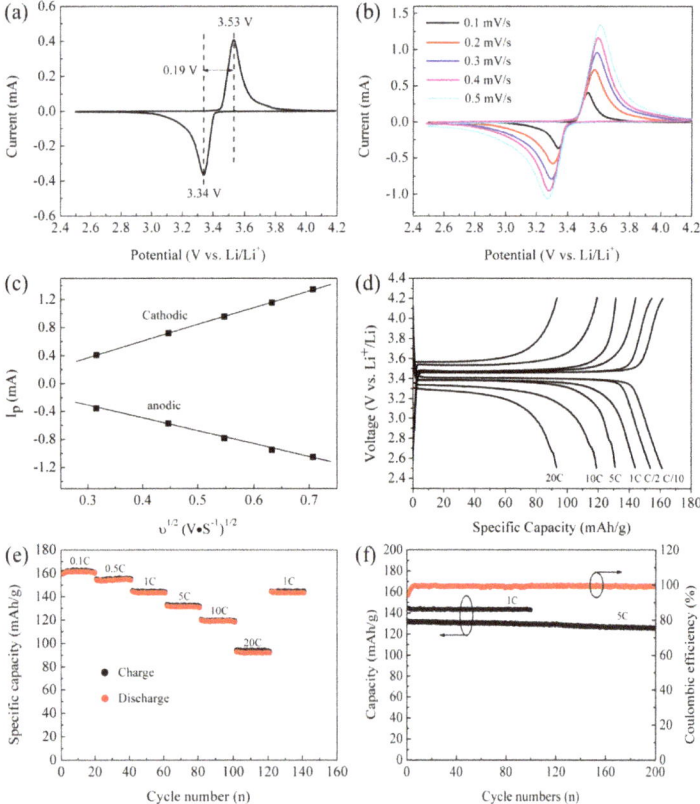

Figure 5. (**a**) Typical CV curve of LiFePO$_4$/C composite at scan rate of 0.1 mV/s; (**b**) CV curves of LiFePO$_4$/C composite at scan rates of 0.1–0.5 mV/s; (**c**) linear response of the peak current (I$_p$) as a function of the square root of scanning rate (ν); (**d**) charge and discharge profiles of LiFePO$_4$/C composite in the potential region from 2.5 to 4.2 V at various rates; (**e**) rate performance curves from 0.1 C to 20 C; and (**f**) cycling performance combined with coulombic efficiency at 1 C and 5 C.

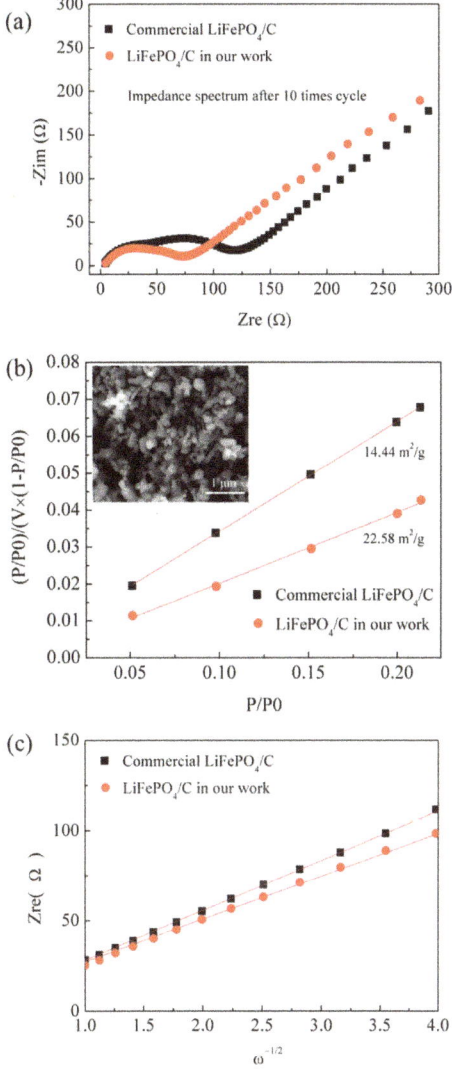

Figure 6. (**a**) The electrochemical impedance spectra (EIS); (**b**) variations and fittings between Z_{Re} and $\omega^{-1/2}$ (the reciprocal square root of the angular frequency ω) in the low-frequency region; and (**c**) specific surface area test (insert is the SEM image of commercial LiFePO$_4$/C) of our LiFePO$_4$/C composite in comparison with those of the commercial LiFePO$_4$/C composite.

4. Conclusions

In conclusion, high-quality LiFePO$_4$/C composite were synthesized via a green route in which no wastewater or air polluting gas is discharged into the environment. The synthesized LiFePO$_4$/C composite exhibited excellent nanoscale particle size (100–200 nm) showing uniform carbon coating on the surface of LiFePO$_4$ nanoparticles, which effectively improved the conductivity and diffusion of Li$^+$ ions of LiFePO$_4$. Consequently, lithium ion batteries using the as-synthesized LiFePO$_4$/C composite as cathode materials exhibit superior electrochemical performance, especially for high rate performance. More importantly, this work provides a valuable method to reduce the manufacturing cost of the

LiFePO$_4$/C cathode materials due to the reduced process for the polluted exhaust purification and wastewater treatment, which is highly desired for applications such as large-scale energy storage and electric vehicles.

Prime Novelty Statement: In this work, we develop a green route to synthesize the LiFePO$_4$/C composite by using Iron (III) oxide (Fe$_2$O$_3$) nanoparticles, Lithium carbonate (Li$_2$CO$_3$), glucose powder and phosphoric acid (H$_3$PO$_4$) solution as raw materials. In the synthesis process, almost no wastewater and air polluted gases are discharged into the environment and the reaction principles are analyzed. The structural, morphological, compositional properties of the LiFePO$_4$/C composite are characterized. Using the LiFePO$_4$/C composite as cathode materials for lithium-ion batteries application, excellent electrochemical performances are obtained, showing a discharge capability of 161 mA h/g at 0.1 C, 119 mA h/g at 10 C and 93 mA h/g at 20 C, and a cycling stability with 98.0% capacity retention at 1 C after 100 cycles and 95.1% at 5 C after 200 cycles. These initial research results are very interesting and the technology developed in this work will provide a valuable approach to reduce the manufacturing cost of LiFePO$_4$/C cathode materials due to the reduced process for the polluted exhaste purification and wastewater treatment.

Author Contributions: Conceptualization, R.L. and J.C.; Methodology, R.L.; Software, Z.L.; Validation, Z.L. and X.A.; Formal Analysis, J.C. and Q.D.; Investigation, Z.L. and X.A.; Resources, R.L.; Data Curation, Z.L.; Writing—Original Draft Preparation, R.L.; Writing—Review & Editing, R.L.; Visualization, Y.P., Z.Z. and M.Y.; Supervision, J.C.; Project Administration, J.C.; Funding Acquisition, Z.Z. and D.F.

Funding: The authors greatly acknowledge the financial support by the Research Program of Shenzhen (JCYJ2017030714570) and the Chinese National Science Foundation (U1601216 and 61505183).

Conflicts of Interest: The authors declare no conflict of interest.

References

1. Yuan, L.X.; Wang, Z.H.; Zhang, W.X.; Hu, X.L.; Chen, J.T.; Huang, Y.H.; Goodenough, J.B. Development and challenges of LiFePO$_4$ cathode material for lithium-ion batteries. *Energy Environ. Sci.* **2011**, *4*, 269–284. [CrossRef]
2. Gong, C.L.; Xue, Z.G.; Wen, S.; Ye, Y.S.; Xie, X.L. Advanced carbon materials/olivine LiFePO$_4$ composites cathode for lithium ion batteries. *J. Power Sources* **2016**, *318*, 93–112. [CrossRef]
3. Wang, Y.; He, P.; Zhou, H. Olivine LiFePO$_4$: development and future. *Energy Environ. Sci.* **2011**, *4*, 805–817. [CrossRef]
4. Liu, J.; Banis, M.N.; Sun, Q.; Lushington, A.; Li, R.; Sham, T.-K.; Sun, X.L. Rational design of atomic-layer-deposited LiFePO$_4$ as a high-performance cathode for lithium-ion batteries. *Adv. Mater.* **2014**, *26*, 6472–6477. [CrossRef] [PubMed]
5. Wang, J.; Sun, X. Olivine LiFePO$_4$: the remaining challenges for future energy storage. *Energy Environ. Sci.* **2015**, *8*, 1110–1138. [CrossRef]
6. Zaghib, K.; Guerfi, A.; Hovington, P.; Vijh, A.; Trudeau, M.; Mauger, A.; Goodenough, J.B.; Julien, C.M. Review and analysis of nanostructured olivine-based lithium rechargeable batteries: Status and trends. *J. Power Sources* **2013**, *232*, 357–369. [CrossRef]
7. Wang, J.; Sun, X. Understanding and recent development of carbon coating on LiFePO$_4$ cathode materials for lithium-ion batteries. *Energy Environ. Sci.* **2012**, *5*, 5163–5185. [CrossRef]
8. Yang, W.Y.; Zhuang, Z.Y.; Chen, X.; Zou, M.Z.; Zhao, G.Y.; Feng, Q.; Li, J.X.; Lin, Y.B.; Huang, Z.G. A simple and novel Si surface modification on LiFePO$_4$@C electrode and its suppression of degradation of lithium ion batteries. *Appl. Surf. Sci.* **2015**, *359*, 875–882. [CrossRef]
9. Zhang, K.; Lee, J.T.; Li, P.; Kang, B.; Kim, J.H.; Yi, G.R.; Park, J.H. Conformal coating strategy comprising N-doped carbon and conventional graphene for achieving ultrahigh power and cyclability of LiFePO$_4$. *Nano Lett.* **2015**, *15*, 6756–6763. [CrossRef] [PubMed]
10. Yang, J.; Wang, J.; Tang, Y.; Wang, D.; Li, X.; Hu, Y.; Li, R.; Liang, G.; Sham, T.K.; Sun, X. LiFePO$_4$–graphene as a superior cathode material for rechargeable lithium batteries: impact of stacked graphene and unfolded graphene. *Energy Environ. Sci.* **2013**, *6*, 1521–1528. [CrossRef]
11. Xu, D.; Chu, X.; He, Y.B.; Ding, Z.; Li, B.; Han, W.; Du, H.; Kang, F. Enhanced performance of interconnected LiFePO$_4$/C microspheres with excellent multiple conductive network and subtle mesoporous structure. *Electrochim. Acta* **2015**, *152*, 398–407. [CrossRef]

12. Gibot, P.; Casas-cabanas, M.; Laffont, L.; Levasseur, S.; Carlach, P.; Hamelet, S.; Tarascon, J.M.; Masquelier, C. Room-temperature single-phase Li insertion/extraction in nanoscale Li$_x$FePO$_4$. *Nat. Mater.* **2008**, *7*, 741–747. [CrossRef] [PubMed]

13. Johnson, I.D.; Lubke, M.; Wu, O.Y.; Makwana, N.M.; Smales, G.J.; Islam, H.U.; Dedigama, R.Y.; Gruar, R.I.; Tighe, C.J.; Scanlon, D.O.; et al. Pilot-scale continuous synthesis of a vanadium-doped LiFePO$_4$/C nanocomposite high-rate cathodes for lithium-ion batteries. *J. Power Sources* **2016**, *302*, 410–418. [CrossRef]

14. Li, Z.; Zhang, D.; Yang, F. Developments of lithium-ion batteries and challenges of LiFePO$_4$ as one promising cathode material. *J. Mater. Sci.* **2009**, *44*, 2435–2443. [CrossRef]

15. Singh, M.; Singh, B.; Willert-Porada, M. Reaction mechanism and morphology of the LiFePO$_4$ materials synthesized by chemical solution deposition and solid-state reaction. *J. Electroanal. Chem.* **2017**, *790*, 11–19. [CrossRef]

16. Cheng, W.H.; Wang, L.; Sun, Z.P.; Wang, Z.J.; Zhang, Q.B.; Lv, D.D.; Ren, W.; Bian, L.; Xu, J.B.; Chang, A.M. Preparation and characterization of LiFePO$_4$·xLi$_3$V$_2$(PO$_4$)$_3$ composites by two-step solid-state reaction method for lithium-ion batteries. *Mater. Lett.* **2017**, *198*, 172–418. [CrossRef]

17. Tang, H.; Si, Y.; Chang, K.; Fu, X.; Li, B.; Shangguan, E.; Chang, Z.; Yuan, X.Z.; Wang, H. Carbon gel assisted low temperature liquid-phase synthesis of C-LiFePO$_4$/graphene layers with high rate and cycle performances. *J. Power Sources* **2015**, *295*, 131–138. [CrossRef]

18. Zhang, Y.T.; Xin, P.Y.; Yao, Q. Electrochemical performance of LiFePO$_4$/C synthesized by sol-gel method as cathode for aqueous lithium ion batteries. *J. Alloys Compd.* **2018**, *741*, 404–408. [CrossRef]

19. Wu, G.; Liu, N.; Gao, X.G.; Tian, X.H.; Zhu, Y.B.; Zhou, K.Y.; Zhu, Q.Y. A hydrothermally synthesized LiFePO$_4$/C composite with superior low-temperature performance and cycle life. *Appl. Surf. Sci.* **2018**, *435*, 1329–1336. [CrossRef]

20. Yen, H.; Rohan, R.; Chiou, C.Y.; Hsieh, C.J.; Bolloju, S.; Li, C.C.; Yang, Y.F.; Ong, C.W.; Lee, J.T. Hierarchy concomitant in situ stable iron(II)–carbon source manipulation using ferrocenecarboxylic acid for hydrothermal synthesis of LiFePO$_4$ as high-capacity battery cathode. *Electrochim. Acta* **2017**, *253*, 227–238. [CrossRef]

21. Guan, X.M.; Li, G.L.; Li, C.Y.; Ren, R.M. Synthesis of porous nano/micro structured LiFePO$_4$/C cathode materials for lithium-ion batteries by spray-drying method. *Trans. Nonferr. Met. Soc. China* **2017**, *27*, 141–147. [CrossRef]

22. Kashi, R.; Khosravi, M.; Mollazadeh, M. Effect of carbon precursor on electrochemical performance of LiFePO$_4$-C nano composite synthesized by ultrasonic spray pyrolysis as cathode active material for Li ion battery. *Mater. Chem. Phys.* **2018**, *203*, 319–332. [CrossRef]

23. Liu, H.W.; Yang, H.M.; Li, J.L. A novel method for preparing LiFePO$_4$ nanorods as a cathode material for lithium-ion power batteries. *Electrochim. Acta* **2010**, *55*, 1626–1629. [CrossRef]

24. Chen, J.J.; Whittingham, M.S. Hydrothermal synthesis of lithium iron phosphate. *Electrochem. Commun.* **2006**, *8*, 855–858. [CrossRef]

25. Ahn, C.W.; Choi, J.J.; Ryu, J.; Hahn, B.D.; Kim, J.W.; Yoon, W.H.; Choi, J.H.; Park, D.S. Microstructure and electrochemical properties of graphite and C-coated LiFePO$_4$ films fabricated by aerosol deposition method for Li ion battery. *Carbon* **2015**, *82*, 135–142. [CrossRef]

26. Belharouak, I.; Johnson, C.; Amine, K. Synthesis and electrochemical analysis of vapor-deposited carbon-coated LiFePO$_4$. *Electrochem. Commun.* **2005**, *7*, 983–988. [CrossRef]

27. Lou, X.M.; Zhang, Y.X. Synthesis of LiFePO$_4$/C cathode materials with both high-rate capability and high tap density for lithium-ion batteries. *J. Mater. Chem.* **2011**, *21*, 4156–4160. [CrossRef]

28. Lu, C.Y.; Rooney, D.W.; Jiang, X.; Sun, W.; Wang, Z.H.; Wang, J.J.; Sun, K.N. Achieving high specific capacity of lithium-ion battery cathodes by modification with "N–Oc" radicals and oxygen-containing functional groups. *J. Mater. Chem. A* **2017**, *5*, 24636–24644. [CrossRef]

29. Tian, R.Y.; Liu, H.Q.; Jiang, Y.; Chen, J.K.; Tan, X.H.; Liu, G.Y.; Zhang, L.N.; Gu, X.H.; Guo, Y.J.; Wang, H.F.; et al. Drastically enhanced high-Rate performance of carbon-coated LiFePO$_4$ nanorods using a green chemical vapor deposition (CVD) method for lithium ion battery: a selective carbon coating process. *ACS Appl. Mater. Interfaces* **2015**, *7*, 11377–11386. [CrossRef] [PubMed]

30. Jin, Y.; Yang, C.P.; Rui, X.H.; Cheng, T.; Chen, C.H. V$_2$O$_3$ modified LiFePO$_4$/C composite with improved electrochemical performance. *J. Power Sources* **2011**, *196*, 5623–5630. [CrossRef]

31. Ma, J.; Li, B.H.; Du, H.D.; Xu, C.J.; Kang, F.Y. Inorganic-based sol–gel synthesis of nano-structured LiFePO$_4$/C composite materials for lithium ion batteries. *J. Solid State Electrochem.* **2012**, *16*, 1353–1362. [CrossRef]
32. Gupta, H.; Kataria, S.; Balo, L.; Singh, V.K.; Singh, S.K.; Tripathi, A.K.; Verma, Y.L.; Singh, R.K. Electrochemical study of Ionic Liquid based polymer electrolyte with graphene oxide coated LiFePO$_4$ cathode for Li battery. *Solid State Ion.* **2018**, *320*, 186–192. [CrossRef]
33. Chen, J.S.; Cheah, Y.L.; Chen, Y.T.; Japaprakash, N.; Madhavi, S.; Yang, Y.H.; Lou, X.W. SnO$_2$ nanoparticles with controlled carbon nanocoating as high-capacity anode materials for lithium-ion batteries. *J. Phys. Chem. C* **2009**, *113*, 20504–20508. [CrossRef]
34. Wang, M.; Yang, Y.; Zhang, Y.X. Synthesis of micro-nano hierarchical structured LiFePO$_4$/C composite with both superior high-rate performance and high tap density. *Nanoscale* **2011**, *3*, 4434–4439. [CrossRef] [PubMed]
35. Mazora, H.; Golodnitsky, D.; Burstein, L.; Gladkich, A.; Peled, E. Electrophoretic deposition of lithium iron phosphate cathode for thin-film 3D-microbatteries. *J. Power Sources* **2012**, *198*, 264–272. [CrossRef]
36. Chen, Y.Q.; Xiang, K.Q.; Zhou, W.; Zhu, Y.R.; Bai, N.B.; Chen, H. LiFePO$_4$/C ultra-thin nano-flakes with ultra-high rate capability and ultra-long cycling life for lithium ion batteries. *J. Alloys Compd.* **2018**, *749*, 1063–1070. [CrossRef]
37. Xiong, W.; Hu, Q.H.; Liu, S.T. A novel and accurate analytical method based on X-ray photoelectron spectroscopy for the quantitative detection of the lithium content in LiFePO$_4$. *Anal. Methods* **2014**, *6*, 5708–5711. [CrossRef]
38. Gao, C.; Zhou, J.; Liu, G.Z.; Wang, L. Lithium-ions diffusion kinetic in LiFePO$_4$/carbon nanoparticles synthesized by microwave plasma chemical vapor deposition for lithium-ion batteries. *Appl. Surf. Sci.* **2018**, *433*, 35–44. [CrossRef]
39. Salah, A.A.; Mauger, A.; Zaghib, K.; Goodenough, J.B.; Ravet, N.; Gauthier, M.; Gendron, F.; Julien, C.M. Reduction Fe^{3+} of impurities in LiFePO$_4$ from pyrolysis of organic precursor used for carbon deposition. *J. Electrochem. Soc.* **2006**, *153*, 1692–1701. [CrossRef]
40. Shiraishi, K.; Dokko, K.; Kanamura, K. Formation of impurities on phospho-olivine LiFePO$_4$ during hydrothermal synthesis. *J. Power Sources* **2005**, *146*, 555–558. [CrossRef]
41. Zhang, Q.; Huang, S.Z.; Jin, J.; Liu, J.; Li, Y.; Wang, H.E.; Chen, L.H.; Wang, B.J.; Su, B.L. Engineering 3D bicontinuous hierarchically macro-mesoporous LiFePO$_4$/C nanocomposite for lithium storage with high rate capability and long cycle stability. *Sci. Rep.* **2016**, *6*, 25942. [CrossRef] [PubMed]
42. Liang, S.Q.; Cao, X.X.; Wang, Y.P.; Hua, Y.; Pan, A.Q.; Cao, G.Z. Uniform 8LiFePO$_4$·Li$_3$V$_2$(PO$_4$)$_3$/C nanoflakes for high-performance Li-ion batteries. *Nano Energy* **2016**, *22*, 48–58. [CrossRef]
43. Huynh, L.T.N.; Tran, T.T.D.; Nguyen, H.H.A.; Nguyen, T.T.T.; Tran, V.M.; Grag, A.; Le, M.L.P. Carbon-coated LiFePO$_4$–carbon nanotube electrodes for high-rate Li-ion battery. *J. Solid State Electrochem.* **2018**, *22*, 2247–2254. [CrossRef]
44. Tang, K.; Yu, X.Q.; Sun, J.P.; Li, H.; Huang, X.J. Kinetic analysis on LiFePO$_4$ thin films by CV, GITT, and EIS. *Electrochim. Acta* **2011**, *56*, 4869–4875. [CrossRef]
45. Xie, J.; Imanishi, N.; Zhang, T.; Hirano, A.; Takeda, Y.; Yamamoto, O. Li-ion diffusion kinetics in LiFePO$_4$ thin film prepared by radio frequency magnetron sputtering. *Electrochim. Acta* **2009**, *54*, 4631–4637. [CrossRef]
46. Wu, X.L.; Guo, Y.G.; Su, J.; Xiong, J.W.; Zhang, Y.L.; Wan, L.J. Carbon-nanotube-decorated nano-LiFePO$_4$@C cathode material with superior high-rate and low-temperature performances for lithium-ion batteries. *Adv. Energy Mater.* **2013**, *3*, 1155–1160. [CrossRef]
47. Qiao, Y.Q.; Feng, W.L.; Li, J.; Shen, T.D. Ultralong cycling stability of carbon-nanotube/LiFePO$_4$ nanocomposites as electrode materials for lithium-ion batteries. *Electrochim. Acta* **2017**, *232*, 323–331. [CrossRef]
48. Schmidt, J.; Chrobak, T.; Ender, M.; Illig, J.; Kiotz, D.; Ivers-Tiffée, E. Studies on LiFePO$_4$ as cathode material using impedance spectroscopy. *J. Power Sources* **2011**, *196*, 5342–5348. [CrossRef]
49. Gong, C.; Xue, Z.; Wang, X.; Zhou, X.; Xie, X.; Mai, Y. Poly(ethylene glycol) grafted multi-walled carbon nanotubes/LiFePO$_4$ composite cathodes for lithium ion batteries. *J. Power Sources* **2014**, *246*, 260–268. [CrossRef]
50. Lei, X.; Zhang, H.; Chen, Y.; Wang, W.; Ye, Y.; Zheng, C.; Deng, P.; Shi, Z. A three-dimensional LiFePO$_4$/carbon nanotubes/graphene composite as a cathode material for lithium-ion batteries with superior high-rate performance. *J. Alloys Compd.* **2015**, *626*, 280–286. [CrossRef]

 © 2018 by the authors. Licensee MDPI, Basel, Switzerland. This article is an open access article distributed under the terms and conditions of the Creative Commons Attribution (CC BY) license (http://creativecommons.org/licenses/by/4.0/).

Article

Simulation-driven Selection of Electrode Materials Based on Mechanical Performance for Lithium-Ion Battery

Abhishek Sarkar, Pranav Shrotriya * and Abhijit Chandra

Department of Mechanical Engineering, Iowa State University, Ames, IA 50011, USA; asarkar@iastate.edu (A.S.); achandra@iastate.edu (A.C.)
* Correspondence: shrotriy@iastate.edu; Tel.: +1-515-294-1423

Received: 19 February 2019; Accepted: 8 March 2019; Published: 12 March 2019

Abstract: Experimental and numerical studies have shown that mechanical loading associated with lithiation/delithiation may limit the useful life of battery electrode materials. The paper presents an approach to parameterize and compare electrode material performance based on mechanical stability. A mathematical model was developed to determine particle deformation and stress fields based upon an elastic-perfectly plastic constitutive response. Mechanical deformation was computed by combining the stress equilibrium equations with the electrochemical diffusion of lithium ions into the electrode particle. The result provided a time developing stress field which shifts from purely elastic to partially plastic deformation as the lithium-ion diffuses into the particle. The model was used to derive five merit indices that parameterize mechanical stability of electrode materials. The merit indices were used to analyze the mechanical stability for the six candidate electrode materials—three for anode materials and three for the cathode material. Finally, the paper suggests ways to improve the mechanical performance of electrode materials and identifies mechanical properties that need to be considered for selection and optimal design of electrode materials.

Keywords: lithium-ion battery; mechanical stability; material index; parametric analysis; elasto-plastic stress

1. Introduction

Increasing energy demand for high energy density storage devices makes lithium-ion batteries a prime source for energy storage [1]. Graphite/lithium cobalt oxide was the first electrode material [2] but rigorous experimental and theoretical study on high capacity and stable electrode materials have allowed lithium batteries to achieve higher energy density, longer cycle life, and safer operation [3]. Electrode performance has been studied through experiments, molecular dynamics simulation and multiphysics modeling [4]. Recent development has focused on finite element modeling for understanding the thermo-mechanical functioning of electrodes and search for newer battery materials of better mechanical and thermal performance at high charging rates [5,6].

The structure and mechanisms governing lithium-ion diffusion and storage vary for different electrode materials. Lithium manganese oxide (LMO) has a cubic spinel structure with the manganese located in the octahedral sites and the lithium ions occupying the tetrahedral sites in a cubic close-packed array of oxygen [7]. LMO gets oxidized in presence of lithium-ion and diffusion process is governed by the chemical potential across the electrodes [8,9]. The insertion of the lithium ions in the vacant octahedral locations produces the Jahn Teller distorted tetragonal phase. The lithium cobalt oxide (LCO) electrode has a α-NaFeO$_2$ layered structure with a cubic close packed array of oxygen with cobalt and lithium ions occupying octahedral sites in alternating layers. Higher voltage applications could cause structural instability within the delithiated layers of LCO [10,11]. Olivine lithium ferrous phosphate (LFP) electrode has an orthorhombic structure with FeO_6 octahedra and PO4 tetrahedra networked in a one-dimensional channel. These channels in LFP provide pathway for lithium to diffuse effectively. Graphite has a layered graphene sheets in either hexagonal (common) or rhombohedral stacking structure with lithium ions diffusing and intercalating between the graphene layers [12]. With development in nanotechnology, silicon nanoparticle-based electrodes have been developed as a promising anode material because of large energy storage capacity (3579 mAh/g) [13]. Silicon in the first cycle reacts with lithium to transform from crystalline to amorphous structure [14] and the amorphous silicon intercalates from then onwards. In the current analysis, amorphous lithiated silicon electrode has been considered. Lithium titanate (LTO) has a spinel structure with no strain during deformation during two-phase lithiation/delithiation process.

Mechanical stability of electrodes is crucial for life prediction of lithium-ion battery. Analytical models predicting the stability of lithium-ion battery electrodes have either assumed a perfectly plastic material or an elastic material [15,16]. Christensen et al modeled lithiation induced stress assuming that stress is a function of the lithium-ion concentration gradient in the particle [17]. Zhang et al [18,19], modeled the radial and hoop stress in ellipsoidal particles considering elastic deformation of lithium manganese oxide spinel. In contrast, perfect plasticity-based models have been developed for anode materials for silicon [15,20]. These models considered pure plasticity due to small impact of the elastic deformation. TEM analysis by Liu et al [21] has shown plastic deformation of 320% by silicon during lithiation. Current research is being focused on novel materials like layered-lithium nickel manganese cobalt electrodes (NMC), which have high capacity (>250 mAh/g) at high discharge potential (3.6–4.5 V) [22]. However, these electrodes are mechanically unstable in high potential domains.

A rigorous mathematical model was developed for evaluation of elasto-plastic stress state in electrode materials. The model was used to study the stress evolution and fracture response in electrode particles of different materials. A set of five merit indices were created which parameterize the materials based on their mechanical performance and fracture stability. A detailed analysis of these indices provided an insight of the material properties useful for a performance boost of the electrode materials. Six electrode materials, three for the cathode and three for the anode, were selected for this study. The results discussed the electrode materials that stand out among the others for having the better mechanical stability and fracture resistance. The paper discusses the crucial material properties which influence the life of a battery, a set of merit indices to evaluate new materials and provides an approach to improve the mechanical performance of lithium-ion battery electrodes.

2. Mathematical and Parametric Analysis

2.1. Mathematical Model

A lithium-ion battery works by the principle of electrochemical diffusion of lithium-ion due to a potential difference between the electrodes. As the cell discharges, the lithium ions diffuse into the cathode from the anode, thereby converting chemical energy into electrical work. Figure 1 and Equations (1) and (2) describe the reaction in any generic lithium transition metal during the discharging process.

$$Li_{1-x}MO_y + xLi^+ + xe^- \xrightarrow{cathode} LiMO_y \quad (1)$$

$$Li_x A \xrightarrow{anode} xLi^+ + xe^- + A \quad (2)$$

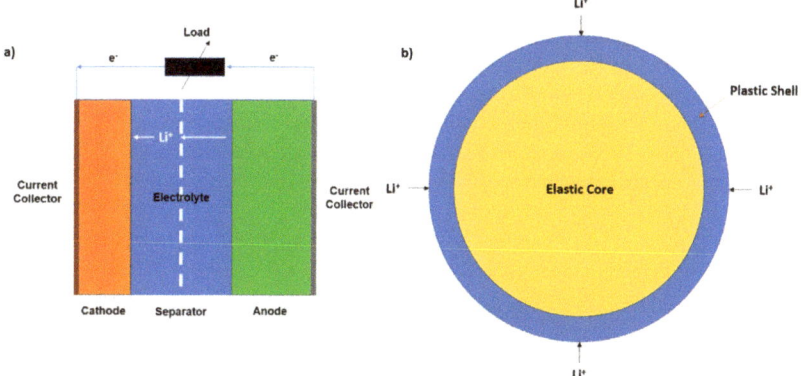

Figure 1. (a) Lithium battery schematic; (b) lithium cathode stress domains; during discharge cycle.

During discharge, lithium ions dissociate from the anode and intercalate with the cathode, made up of a metal oxide of lithium, to form a lithium intercalation compound and vice-versa during charging. Many mathematical models on the elastic deformation of electrode particle predicted the mechanical performance of certain electrode materials [18,19,23]. In the present work, the deformation of the electrode particle in the plastic regime was modeled. The porous electrode theory [17] with the electrolyte having an infinite source of lithium ion, free expansion of the particle surface and the perfectly plasticity of the material at yield point were assumed for modeling. The mass transport of lithium ions in the electrode material is expressed below where M is the mobility and c is the concentration of lithium-ion.

$$J = -cM\nabla \mu \tag{3}$$

The electrochemical potential (μ) is expressed as a function of lithium-ion mole fraction (x) and the chemical stress generated.

$$\mu = \mu^0 + RT\ln(x) - \Omega\sigma_h \tag{4}$$

σ_h is the hydrostatic stress on the particle due to differential expansion, Ω is the partial molar volume, and x is the mole fraction of lithium in the particle. Substituting the electrochemical potential (Equation (4)) into mass flux (Equation (3)), the equivalent stress dependent ion flux equation was derived.

$$J = -Mc\left[RT\frac{\nabla c}{c} - \Omega\nabla\sigma_h\right] = -D\left[\nabla c - \frac{\Omega c}{RT}\nabla\sigma_h\right] \tag{5}$$

D is the mass diffusion coefficient of the lithium-ion in the electrode, and T is the operating temperature. Fick's Law of diffusion was expressed for a spherically symmetric particle from the diffusion mass flux (Equation (5)).

$$\frac{\partial c}{\partial t} = D\left[\frac{1}{r^2}\frac{\partial}{\partial r}\left(r^2\frac{\partial c}{\partial r}\right) - \frac{\Omega}{RT}\left\{\frac{\partial c}{\partial r}\frac{\partial \sigma_h}{\partial r} + \frac{c}{r^2}\frac{\partial}{\partial r}\left(r^2\frac{\partial \sigma_h}{\partial r}\right)\right\}\right] \tag{6}$$

The following boundary conditions were used to solve Equation (6).

$$i = \frac{\alpha\rho r_0}{3}C_{rate} \tag{7}$$

$$J|_{r=r_0} = \frac{i}{F} \tag{8}$$

$$\left.\frac{\partial c}{\partial r}\right|_{r=0} = 0 \tag{9}$$

Equation (7) shows the current flux (*i*) depends on the theoretical capacity of the electrode material (α), density (ρ), radius (r_0) and the rate of charging (C_{rate}).

Equation (6) shows hydrostatic stress coupled with lithium-ion concentration, which affects the diffusion process. The intercalation of lithium-ion into the electrode causes an expansion of the particle. Particle expansion leads to shifting of atomic planes causing dislocation of atoms from their natural lattice sites. This leads to the generation of a stress field in the particle. Therefore, the formulation of the elastic component of strain was found similar to the strain produced during to thermal expansion [18].

$$\varepsilon^e_r = \frac{1}{E}[\sigma_r - 2\nu\sigma_\theta] + \frac{\tilde{c}\Omega}{3} = \frac{du}{dr} \tag{10}$$

$$\varepsilon^e_\theta = \frac{1}{E}[(1-\nu)\sigma_r - \nu\sigma_\theta] + \frac{\tilde{c}\Omega}{3} = \frac{u}{r} \tag{11}$$

$$\tilde{c} = c - c_0 \tag{12}$$

The radial (ε_r) and hoop (ε_θ) strain depended upon the concentration variation across the particle. *u* is the radial displacement and c_0 is the initial lithium-ion concentration. As the electrode was further lithiated, the equivalent stress exceeded the yield stress causing the material to deform plastically. For a spherical particle, the yield criterion follows the condition below.

$$|\sigma_r - \sigma_\theta| \leq S_y \tag{13}$$

When the equivalent stress exceeds the yield limit (S_y), the total strain generated was due to both the elastic and plastic deformation of the particle. Figure 1b represents the elasto-plastic schematic of a cathode particle during lithiation. The total (hydrostatic) strain would be equal to the summation of the volumetric elastic and plastic strain.

$$\varepsilon_{ij} = \varepsilon^e_{ij} + \varepsilon^p_{ij} \tag{14}$$

Since the particle was assumed completely spherical, the stress equilibrium equation was solved in the spherical coordinates considering radial symmetry.

$$\frac{\partial \sigma_r}{\partial r} + \frac{2}{r}(\sigma_r - \sigma_\theta) = 0 \tag{15}$$

σ_r is the radial stress component and σ_θ is the hoop stress component. This equation was solved for the elastic segment by substituting the elastic displacements and for the plastic part by substituting the yield equality using the following boundary conditions [24].

$$\left.\frac{\partial \sigma^{el}_r}{\partial r}\right|_{r=0} = 0 \tag{16}$$

$$\sigma^{el}_r = \sigma^{pl}_r|_{r_p} \tag{17}$$

These conditions satisfied the radial stress continuum at the core and interface between elastic and plastic domain (r_p). The plastic stress was solved by substituting the yield equality into the stress equilibrium equation with free expansion along the radial direction on the surface.

$$\sigma^{pl}_r|_{r_o} = 0 \tag{18}$$

The stress (radial and hoop) was calculated by solving the elastic-perfectly plastic equations for the plastic domain.

$$\sigma^{pl}_r = 2S_y \ln\left[\frac{r_o}{r}\right]; r_p \leq r \leq r_o \tag{19}$$

$$\sigma_\theta^{pl} = 2S_y \ln\left[\frac{r_o}{r}\right] - Y; r_p \leq r \leq r_o \tag{20}$$

The differential equation for the elastic domain was solved by using the plastic radial stress at the plastic interface.

$$\sigma_r^{el} = 2S_y \ln\left[\frac{r_o}{r_p}\right] + \frac{2\Omega E}{3(1-\nu)}\left[\frac{1}{r_p^3}\int_0^{r_p}\tilde{c}r^2 dr - \frac{1}{r^3}\int_0^r \tilde{c}r^2 dr\right]; 0 \leq r \leq r_p \tag{21}$$

$$\sigma_\theta^{el} = 2S_y \ln\left[\frac{r_o}{r_p}\right] + \frac{\Omega E}{3(1-\nu)}\left[\frac{2}{r_p^3}\int_0^{r_p}\tilde{c}r^2 dr + \frac{1}{r^3}\int_0^r \tilde{c}r^2 dr - \tilde{c}\right]; 0 \leq r \leq r_p \tag{22}$$

The mean (hydrostatic) stress was then found for the elastic and plastic equations.

$$\sigma_h = \frac{\sigma_r + 2\sigma_\theta}{3} \tag{23}$$

$$\sigma_h^{pl} = 2S_y \ln\left[\frac{r_o}{r}\right] - \frac{2}{3}Y; r_p \leq r \leq r_o \tag{24}$$

$$\sigma_h^{el} = 2S_y \ln\left[\frac{r_o}{r_p}\right] + \frac{2\Omega E}{9(1-\nu)}\left[\frac{3}{r_p^3}\int_0^{r_p}\tilde{c}r^2 dr - \tilde{c}\right]; 0 \leq r \leq r_p \tag{25}$$

The elastic and plastic stresses in Equations (24) and (25) were substituted into Equation (6) to decouple the concentration from stress.

$$\frac{\partial c}{\partial t} = D\left[\frac{1}{r^2}\frac{\partial}{\partial r}\left(r^2\frac{\partial c}{\partial r}\right) + \theta\left(\frac{\partial c}{\partial r}\right)^2 + \theta c\left\{\frac{1}{r^2}\frac{\partial}{\partial r}\left(r^2\frac{\partial c}{\partial r}\right)\right\}\right]; 0 \leq r \leq r_p \tag{26}$$

$$\frac{\partial c}{\partial t} = D\left[\frac{1}{r^2}\frac{\partial}{\partial r}\left(r^2\frac{\partial c}{\partial r}\right) - \frac{\Pi}{r}\left(\frac{c}{r} + \frac{\partial c}{\partial r}\right)\right]; r_p \leq r \leq r_o \tag{27}$$

$\theta = \frac{2\Omega^2 E}{9RT(1-\nu)}$ and $\Pi = \frac{2S_y\Omega}{RT}$ are constants which coupled the gap between stress driven concentration and diffusion driven concentration in the elastic and plastic equations, respectively.

Although fatigue failure of lithium-ion electrode is a more dominant mechanism than fracture, but the ability of the material to prevent crack propagation is a primary characteristics needed to avoid the onset of failure. A simple fracture analysis was used to compare different electrode materials based on their material toughness and ability to prevent crack propagation upon loading. The fracture of the spherical electrode during lithiation occurs due to the hoop stress. The simplest model for an edge crack relates the stress intensity with the applied stress and crack size.

$$K_I = C\sigma_\theta\sqrt{\pi a} \tag{28}$$

a is the crack depth from the surface, K_I is the stress intensity factor and C is a geometric factor.

2.2. Material Characterization for Lithium Electrodes

The life of the battery depends on several parameters of which the hoop stress and critical stress intensity of the given material are crucial. The diffusion-induced stress is governed by physical and material properties like particle diameter, state of charge, theoretical capacity, specific molar volume, yield strength, Young's Modulus, etc. Since the lithiation process is diffusive, it is essential to compare different candidate electrode materials based on their material properties as recorded in Table 1.

In the present work, the parametric analysis was performed based on the selection of a quantity (parameter) that needed to be optimized. A set of constraints and a free variable were used to substitute the constraint in the parameter for optimization [25]. Five material indices (M) were created to express mechanical performance and fracture resistance.

Table 1. Material properties of selected electrode materials.

Properties	LiMn$_2$O$_4$	LiCoO$_2$	LiFePO$_4$	Li$_x$C$_6$	Li$_x$Si$_{15}$	Li$_x$TiO$_2$
D (m^2/s)	7.08 × 10^{-15} [17]	1.00 × 10^{-13} [26]	7.96 × 10^{-16} [27]	3.90 × 10^{-14} [26]	1.00 × 10^{-16} [28]	6.80 × 10^{-15} [29]
ρ (kg/m)	4100 [17]	5030 [26]	3600 [30]	2100 [26]	2328 [31]	3510 [32]
a_{th} (mAh/g)	148 [33]	166 [33]	170 [33]	372 [33]	4200 [33]	175 [7]
c_{max} (mol/m^3)	2.29 × 10^4 [17]	4.99 × 10^4 [26]	2.12 × 10^4	3.05 × 10^4 [26]	8.87 × 10^4 [34]	5.00 × 10^4 [29]
Ω (m^3/mol)	3.50 × 10^{-6} [17]	1.92 × 10^{-6} [26]	67.32 × 10^{-6} [35]	3.17 × 10^{-6} [26]	32.25 × 10^{-6} [36]	5.00 × 10^{-6} [29]
S_y (MPa)	776 [37]	1056 [38]	500 [38]	23 [33]	720 [21]	836
E (GPa)	194 [38]	264 [38]	125 [38]	10 [38]	12 [39]	209 [38]
N	0.26 [38]	0.32 [38]	0.28 [38]	0.24 [38]	0.25 [38]	0.19 [38]
K_{1C} $(MPam^{0.5})$	1.50 [40]	1.30 [40]	1.50 [40]	1.25 [40]	1.00 [40]	1.50 [40]

Constraints and Free Variables:

In the mechanical parametric analysis of battery systems, it is important to set certain constraints to limit the degree of variability of the material indices for comparison of materials. Setting the average non-dimensional concentration profile to a constant allows all electrode particles to have the same amount of lithium ions diffused within them.

$$\hat{c}(\hat{r}) = \frac{c}{c_{max}} = Cons \qquad (29)$$

Setting time of diffusion or the particle radius as constant became slightly challenging as keeping equal radius particle seems to be a more apt decision from a manufacturing perspective. However, the time required for charging was crucial compared to the radius of the particle. Comparison based on equality of temporal coordinates was found more suitable in this situation.

$$\hat{t} = \frac{tD}{r_0^2} \qquad (30)$$

Considering the total time required for diffusion was kept constant, the following could be derived.

$$r_0 \propto \sqrt{D} \qquad (31)$$

The radius emerges out to be the free variable which substitutes the diffusion coefficient in the material indices. Another constraint was obtained from the yield criterion.

$$\sigma_r^{el} - \sigma_\theta^{el} = \frac{\Omega E}{3(1-\nu)} \left[\tilde{c} - \frac{3}{r_p^3} \int_0^{r_p} \tilde{c} r^2 dr \right] \leq S_y \qquad (32)$$

Since the integral term in Equation (32) became constant while substituting $\tilde{c}$ from Equation (29), the maximum concentration allowed became proportional to the yield stress.

$$c_{max} \propto \frac{S_y(1-\nu)}{\Omega E} \qquad (33)$$

For the fracture analysis, it was considered that the upper limit for the stress intensity factor was bound by the fracture toughness of the material.

$$K_I \leq K_{1C} \qquad (34)$$

3. Results and Discussion

During lithiation, the lithium-ion concentration is higher near the surface of the particle and decreases near the core which occurs due to the effect of Fick's law of mass diffusion. The intercalation of lithium-ion with the electrode material causes it to expand proportionally to the relative concentration of lithium-ion. The surface of the particle tries to expand more during lithiation than the core. Expansion causes the surface to be under compression while the core remains under tension. During delithiation, the surface lithium-ion concentration is lower than the core. Hence, the surface contracts faster causing tension on the surface and compression in the core. In real world application, the deformation of the electrode material could be detected during testing and control stage of manufacturing or in development of newer electrode materials. The strain response of the electrode would vary under different rates of charging. The stress thus induced in the electrode to oppose this deformation would cause cracks to propagate and the electrode to fail. Electrode materials with strain exceeding the elastic limit would deform plastically. Under such loading conditions, the material would not relax back to its original shape upon unloading. This could be severely problematic for electrodes with higher elastic modulus, as cyclic shape deformation and high residual stress would lead to faster fatigue failure. On the other hand, elastic loading is commonly observed in electrode particles upon lithiation, where the strain is below the critical limit. An elastic-perfectly plastic chemical diffusion model was developed to perform the stress and fracture analysis during lithiation of different electrode materials. The electrode materials were compared based on their mechanical stability, ability to handle faster charging without yielding and higher fracture characteristics. The lithiation process was considered without effects of reaction or phase transformation. The merit indices were developed considering constant state of charge, charging time and temperature of 298 K. The equations for the stress analysis and merit indices were solved using MATLAB platform.

The mathematical analysis was done considering the electrode material to be elastic and perfectly plastic. Therefore, when the yield criteria (Equation (13)) was reached, the equivalent stress ($|\sigma_r - \sigma_\theta|$) remained equal to the yield stress of the material, while the material kept expanding plastically. The plastic deformability was experimentally observed by Kosova et al [41] and Schilcher et al [42]. In Figure 2, the normalized stress distribution was plotted against the normalized radius of a lithium manganese oxide particle. The particle was considered of 10μm in radius and charged under 2C and 3C rates of charging. For the 2C charging rate, the lithium manganese oxide particle was found barely plastic near the particle surface, where the equivalent stress equaled the yield stress of the material. However, with an increase in the charging rate to 3C, the electrode particle deformed plastically from the surface till about $0.65r_0$. The stress distribution in 2C case was found to be much more uniform, while for higher rate of charging, the stress profile became sharp in the plastic shell. This pushed the tensile stress domain in the core to a very high stress state. It is interesting to note that the stress (radial and hoop) developed 1.3 times and 2.4 times in magnitude of the yield stress for the 2C and 3C cases, respectively. However, the material did not fail under such high loads because the equivalent stress near the core was nearly zero. Hence, the material was under a purely tensile hydrostatic load which prevented failure because the electrode particle was considered to be solid without any crack. It could be inferred that the presence of small voids or microcracks in this domain, as observed experimentally in silicon [37], would lead to the voids to coalesce and it would form cracks. These cracks would propagate rapidly towards the surface and would get closed in the compressive domain near the surface. This would lead to failure of the electrode material above a certain dimension and rate of charging. It could then be inferred that lithium manganese oxide particles of 10μm radius are safe for operation under 2C charging rate.

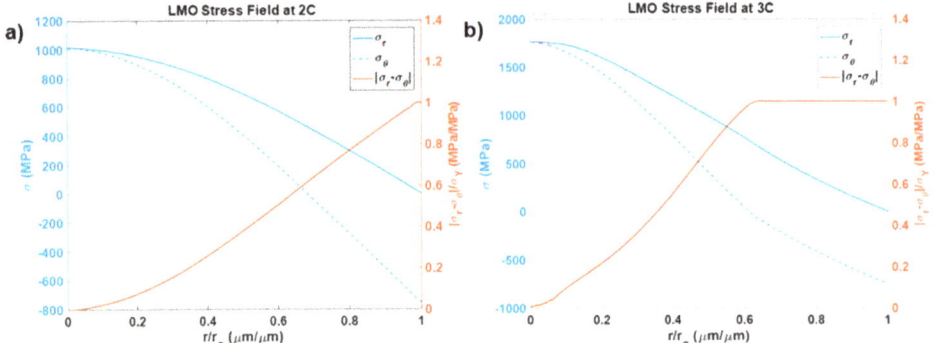

Figure 2. Stress distribution profile of a spherical lithium manganese oxide particle during lithiation under rates of charging of (**a**) 2C, and (**b**) 3C.

3.1. Charging Index

Faster rate of charging is one of the most desired outcomes of battery research. Under the constraint of yield (Equation (33)), the boundary condition in Equation (8) was modified by substituting the flux from Equation (5) and elastic stress from Equation (25).

$$-D\frac{\partial c}{\partial r}(1+\theta c) = \frac{i}{F} \qquad (35)$$

Equation (35) was normalized with $\hat{c} = \frac{\tilde{c}}{c_{max}}$, $\hat{r} = \frac{r}{r_o}$ and the current flux term was expanded from Equation (7).

$$-\frac{\partial \hat{c}}{\partial r}(1+\theta c_{max}\hat{c}) = \frac{\rho \alpha r_o^2}{DFc_{max}}C_{rate} \qquad (36)$$

Ignoring one in the left bracket and considering constraint of constant concentration and gradient from Equation (29), time constraint (Equation (30)) and approximation for maximum concentration from Equation (33), the rate of charging could be evaluated. The following merit index was minimized for materials that handles higher rates of charging.

$$M_{Cr} = \frac{\rho E \alpha}{S_y^2(1-\nu)^2} \qquad (37)$$

The charging rate merit index parameterized materials based on their ability to handle high charging rates without yielding. This index was developed with the constraint on the maximum lithium ion concentration that the electrode material could hold. Moreover, the time of complete lithiation remained constant for all the materials. A higher charging index indicated that the material that can store more charge in a lesser time without yielding, shows better promise as a future battery material for HEV battery systems.

3.2. Elastic and Plastic Indices

The differential equations (Equations (26) and (27)) were related to stress-based diffusion through θ and Π which influenced the diffusion process. The elastic equation (Equation (26)) was promoted by θ (additive), while the plastic equation (Equation (27)) reduced with the increase of Π (subtractive). For lower stress response, it could be concluded that the concentration gradient during diffusion should be minimized for minimization of the following elastic merit index and the plastic merit index.

$$M_{El} = \frac{\Omega^2 E c_{max}}{(1-\nu)} \qquad (38)$$

$$M_{Pl} = \frac{1}{S_y \Omega} \qquad (39)$$

The diffusion equation (Equation (25)) got modified by the addition of extra concentration terms from the mean elastic stress field. These terms deviated the concentration distribution from its general parabolic structure to a higher gradient distribution, resulting in hindered diffusion in the particle. Higher gradient, or slower diffusion, therefore, leads to large stress build up and battery not being utilized to its utmost potential. It is crucial to understand the importance of this effect and to find ways to minimize its impact on the diffusion process.

3.3. Stress Index

Under elastic loading, the hydrostatic stress in Equation (25) was influenced by the concentration gradient in the particle. For small plastic stress effects, the equation was simplified.

$$\hat{\sigma}_h^{el} \propto \frac{\Omega E c_{max}}{S_y(1-\nu)} \left[\frac{3}{\hat{r}_p^3} \int_0^{\hat{r}_p} \hat{c}\hat{r}^2 dr - \bar{\hat{c}} \right] \qquad (40)$$

The stress was normalized by the yield stress and the integral part of the equation was considered constant (Equation 29 in 40). The following merit index for stress was minimized for prediction of low stress-induced materials.

$$M_{St} = \frac{\Omega E c_{max}}{S_y(1-\nu)} \qquad (41)$$

3.4. Fracture Index

The fracture formulation (Equation (28)), with the hoop stress from Equation (25) (simplified for negligible yield stress) and the fracture toughness constraint (Equation (34)) was used to express the detectable crack length, which needed to be maximized for longer life. Therefore, the merit index was minimized in order to select a material with high fracture resistance.

$$M_{Fr} = \frac{\Omega E c_{max}}{K_{1C}(1-\nu)} \qquad (42)$$

Material under stress tends to fail in the presence of cracks. Cracks or flaws act as stress concentrators which magnify the local stress field. If the crack size is beyond the critical limit, the stress at the crack tip exceeds the fracture strength of the material causing failure due to crack propagation.

Figure 3 demonstrated a multivariable comparison between the merit index based on maximizing the charging rate and the merit index for reducing the effect of elastic stress on diffusion. The objective of this comparison was selection of high capacity electrode materials to maintain lithium-ion storage capacity under high elastic deformation. The elastic merit index was dependent on the partial molar volume, modulus of elasticity and maximum concentration of lithium that the material can store. All these parameters were minimized for lower elastic effects on diffusion.

The selected electrode materials, three for the cathode and three for the anode, were compared based on the formulated indices. Figure 3 shows that lithium manganese oxide was the most suitable material for the cathode and lithium titanate showed good promise for anode based on their high charge storage capability and low effects of elastic stress on lithium diffusion. Lithium manganese oxide was found experimentally more stable than commercially used lithium cobalt oxide [43]. The three-dimensional structure of the manganese oxide spinel allowed more space for intercalation with lithium ions during discharge and vice-versa for charge [9]. This means that manganese oxide allows faster lithiation without significant deformation. Silicon showed a tremendous performance compared to graphite, based on the ability to be operated at high charging rates [44]. However, graphite being soft generated lower elastic stresses during lithiation. While silicon did not perform well because of its high molar volume. Silicon expanded by 4 times its volume on lithiation leading to a severe effect on lithium diffusion during lithiation. Lithium titanate showed good performance amongst the anode material due to its higher yield strength and high capacity. The compared based on the ability to store charge silicon exceeded lithium titanate but its lower yield strength made it more prone to yielding than the later.

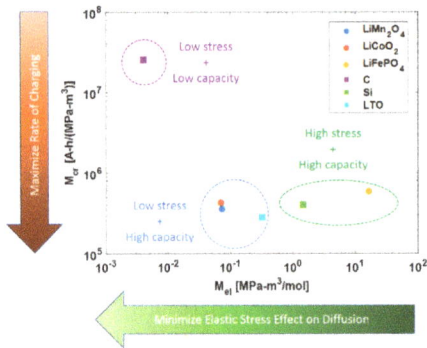

Figure 3. Material Selection based on Charging Rate Index and Elastic Stress Index.

The charging index depends directly on the square of the yield strength of the material and inversely to the charge capacity and elastic modulus. Materials like lithium ferrous phosphate have a lower rating on this index. It is advisable to apply strengthening mechanism to improve the yield strength. Furthermore, lithium ferrous phosphate showed very poor characteristics on the elastic merit index scale because of its very high molar volume. This meant that ferrous phosphate electrodes deformed elastically more than lithium manganese oxide and lithium cobalt oxide electrodes. Newer materials could be hardened and/or alloyed to improve their performance. This would allow materials to be charged quicker without failure. Furthermore, the stress developed in the particle came from the mass flux of lithium ions. It was noted from Equation (7) that if the radius of the particle was reduced, it would allow higher charging rates for same flux. However, this is a tradeoff between manufacturability of smaller particles against the mechanical performance of the electrode.

As observed, faster charging increased the slope of the concentration profile in the electrode particle. It caused the equivalent stress of the particle to exceed the yield limits and the particle deforms plastically. Plastic mean stress also affected the diffusion process. However, the yield stress and molar volume reduced the concentration gradient and allowed free expansion of the particle. Therefore, it is required to maximize this scale for minimization of plastic deformation. Figure 4 compared the materials by maximizing the merit index for charging while minimizing the plastic deformation effect on diffusion. Lithium ferrous phosphate electrode was an excellent cathode material under this category in comparison to lithium manganese oxide and cobalt oxide. The merit index showed good the performance of silicon for the anode. Silicon's ability to be lithiated under plastic deformation was observed experimentally [21], which validated the integrity of the plastic merit index. Graphite showed very poor plastic performance because of its low yield strength and brittle nature, making it vulnerable in the plastic deformation domain. Therefore, silicon is an excellent choice for HEV batteries which has very high capacities with the ability to operate under plastic deformation.

The strength of the material was compared against the material toughness. Figure 5 shows a comparison between the diffusion induced hydrostatic stress index against the fracture resistance index for different electrode materials. It could be inferred that higher yield strength means better mechanical performance, lower elastic modulus means a softer material which generates less stress during deformation, while lower molar volume means smaller volumetric deformation during lithiation. The study also showed that lithium manganese oxide showed the best mechanical performance as the cathode. The stress developed due to the concentration gradient, which is the prime source of stress, was minimum for lithium manganese oxide and was the highest for lithium ferrous phosphate. This was attributed due to the lower modulus of elasticity and molar volume of manganese oxide, which developed low stress and high fracture toughness, making it more resilient to fracture. Silicon was found comparably poor to graphite because of its low elastic modulus. However, silicon was comparable to other cathodic material making it the second-best candidate. Silicon can store more charge than graphite making this tradeoff favorable towards the former. To improve mechanical characteristics of new materials, it is important to reduce the molar volume and/or increase the yield strength.

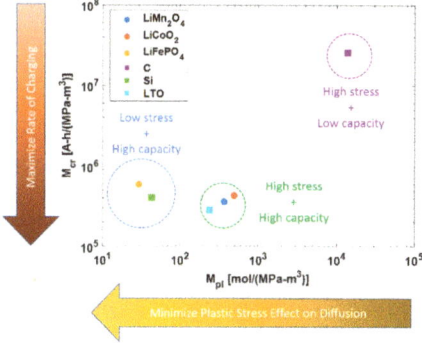

Figure 4. Material Selection based on Charging Rate Index and Plastic Stress Index.

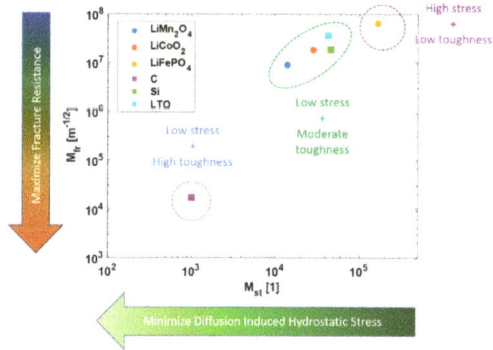

Figure 5. Material Selection based on Fracture Index and Hydrostatic Stress Index.

Different chemo-mechanical processes can harden the material making it durable. Doping the electrode material with a chemical reagent is an excellent way to alter the molar volume of the material. Alteration of the stoichiometry of the material reduces the distortion strains resulting in less expansion and improvement in fracture characteristics like toughness etc. Silicon, for example, can be made tougher by coating it with a more resilient inert layer like titanium oxide [45]. The selection of material under high charging rates depends on their mechanical performance and fracture stability. Higher mechanical stability of material having higher yield strength can be achieved by toughening the material. Size of the electrode particle also plays a crucial role in determining its performance. Smaller particles can be charged faster for the same stress and they allow better diffusion. It also leads to the lower concentration gradient of lithium-ion and consequently better mechanical stress characteristics. Silicon has a very high charge capacity which makes it a good choice for high charging systems where plastic deformation is dominant. However, its fracture characteristics need to be improved for a safer design.

4. Conclusions

The paper discussed an elasto-plastic based diffusion-induced stress model and a set of five material indices for material categorization and selection. The model was used to determine the concentration profile for lithium-ion in the elastic and plastic domain. The concentration was used to find the stress profile for the electrode during lithiation. The comparison between the stress profile for lithium manganese oxide under 2C and 3C charging rates showed that the center of the particle was under high tensile hydrostatic loading during lithiation. The equivalent stress being zero at the core prevents failure of the particle. However, the presence of voids would lead to failure by crack nucleation and propagation over multiple cycles. The materials were evaluated based on their charge holding capability under elastic and plastic loading. It was found that under elastic loading conditions, lithium manganese oxide and graphite are the best cathode and anode materials while under plastic loading, lithium ferrous phosphate and silicon are the best material choice for the battery. The strength of the electrode material was compared against its toughness, and lithium manganese oxide and graphite showed the best performance. From the material parameterization, it was inferred that lithium manganese oxide was the most suited cathode material due to its ability to perform well under faster charging and showing excellent mechanical and fracture characteristics. Graphite performed great in handling elastic stress and was resistant to fracture. However, newer materials like silicon and lithium titanate were found to be good for faster charging and in handling plastic deformation. This made them good choices for HEV battery modules. Furthermore, newer materials could be made better by lowering the elastic modulus and molar volume and improving the yield strength by toughening. Fracture toughness could be improved by coating the material with a more resilient substance to absorb the fracture energy. Reducing particle size would be a great alternative, as it would allow higher rates of charging for same current flux and lower mechanical stresses. These indices have a great importance in classifying good electrode materials and aid in the search for newer battery materials.

Author Contributions: Conceptualization, A.S., P.S. and A.C.; methodology, A.S.; software, A.S.; validation A.S., P.S. and A.C.; formal analysis, A.S.; investigation, A.S., P.S. and A.C.; resources, A.C. and P.S.; data curation, A.S.; writing—original draft preparation, A.S.; writing—review and editing, P.S. and A.S.; visualization, A.S.; supervision, P.S. and A.C.; project administration, P.S. and A.C.

Funding: This research received no external funding.

Conflicts of Interest: The authors declare no conflict of interest.

References

1. Armand, M.; Tarascon, J.-M. Building better batteries. *Nature* **2008**, *451*, 652. [CrossRef] [PubMed]
2. Whittingham, M.S. Lithium batteries and cathode materials. *Chem. Rev.* **2004**, *104*, 4271–4302. [CrossRef] [PubMed]
3. Ellis, B.L.; Lee, K.T.; Nazar, L.F. Positive electrode materials for li-ion and li-batteries. *Chem. Mater.* **2010**, *22*, 691–714. [CrossRef]
4. Lu, L.; Han, X.; Li, J.; Hua, J.; Ouyang, M. A review on the key issues for lithium-ion battery management in electric vehicles. *J. Power Sources* **2013**, *226*, 272–288. [CrossRef]
5. Ramadesigan, V.; Northrop, P.W.C.; De, S.; Santhanagopalan, S.; Braatz, R.D.; Subramanian, V.R. Modeling and simulation of lithium-ion batteries from a systems engineering perspective. *J. Electrochem. Soc.* **2012**, *159*, R31–R45. [CrossRef]
6. Kermani, G.; Sahraei, E. Review: Characterization and modeling of the mechanical properties of lithium-ion batteries. *Energies* **2017**, *10*, 1730. [CrossRef]
7. Xu, J.J.; Kinser, A.J.; Owens, B.B.; Smyrl, W.H. Amorphous manganese dioxide: A high capacity lithium intercalation host. *Electrochem. Solid State Lett.* **1998**, *1*, 1–3. [CrossRef]

8. Paulsen, J.M.; Dahn, J.R. Phase diagram of Li-Mn-O spinel in air. *Chem. Mater.* **1999**, *11*, 3065–3079. [CrossRef]
9. Gabrisch, H.; Yazami, R.; Fultz, B. Hexagonal to cubic spinel transformation in lithiated cobalt oxide TEM investigation. *J. Electrochem. Soc.* **2004**, *151*, A891–A897. [CrossRef]
10. Padhi, A.K.; Nanjundaswamy, K.S.; Goodenough, J.B. Phospho-olivines as positive-electrode materials for rechargeable lithium batteries. *J. Electrochem. Soc.* **1997**, *144*, 1188–1194. [CrossRef]
11. Etacheri, V.; Marom, R.; Elazari, R.; Salitra, G.; Aurbach, D. Challenges in the development of advanced li-ion batteries: A review. *Energy Environ. Sci.* **2011**, *4*, 3243–3262. [CrossRef]
12. Chang, J.; Huang, X.; Zhou, G.; Cui, S.; Hallac, P.B.; Jiang, J.; Hurley, P.T.; Chen, J. Lithium-ion batteries: Multilayered si nanoparticle/reduced graphene oxide hybrid as a high-performance lithium-ion battery anode. *Adv. Mater.* **2014**, *26*, 665. [CrossRef]
13. Wang, F.; Wu, L.; Key, B.; Yang, X.-Q.; Grey, C.P.; Zhu, Y.; Graetz, J. Electrochemical reaction of lithium with nanostructured silicon anodes: A study by in-situ synchrotron X-Ray diffraction and electron energy-loss spectroscopy. *Adv. Energy Mater.* **2013**, *3*, 1324–1331. [CrossRef]
14. Zhao, K.; Pharr, M.; Wan, Q.; Wang, W.L.; Kaxiras, E.; Vlassak, J.J.; Suo, Z. Concurrent reaction and plasticity during initial lithiation of crystalline silicon in lithium-ion batteries. *J. Electrochem. Soc.* **2012**, *159*, A238–A243. [CrossRef]
15. Pharr, M.; Zhao, K.; Wang, X.; Suo, Z.; Vlassak, J.J. Kinetics of initial lithiation of crystalline silicon electrodes of lithium-ion batteries. *Nano Lett.* **2012**, *12*, 5039–5047. [CrossRef] [PubMed]
16. Christensen, J.; Newman, J. A mathematical model of stress generation and fracture in lithium manganese oxide. *J. Electrochem. Soc.* **2006**, *153*, A1019–A1030. [CrossRef]
17. Zhang, X.; Shyy, W.; Sastry, A.M. Numerical simulation of intercalation-induced stress in li-ion battery electrode particles. *J. Electrochem. Soc.* **2007**, *154*, A910–A916. [CrossRef]
18. Zhang, X.; Sastry, A.M.; Shyy, W. Intercalation-induced stress and heat generation within single lithium-ion battery cathode particles. *J. Electrochem. Soc.* **2008**, *155*, A542–A552. [CrossRef]
19. Sethuraman, V.A.; Nguyen, A.; Chon, M.J.; Nadimpalli, S.P.V.; Wang, H.; Abraham, D.P.; Bower, A.F.; Shenoy, V.B.; Guduru, P.R. Stress evolution in composite silicon electrodes during lithiation/delithiation. *J. Electrochem. Soc.* **2013**, *160*, A739–A746. [CrossRef]
20. Liu, X.H.; Zhong, L.; Huang, S.; Mao, S.X.; Zhu, T.; Huang, J.Y. Size-dependent fracture of silicon nanoparticles during lithiation. *ACS Nano* **2012**, *6*, 1522–1531. [CrossRef]
21. Xiang, X.; Knight, J.C.; Li, W.; Manthiram, A. Understanding the effect of Co_3+ substitution on the electrochemical properties of lithium-rich layered oxide cathodes for lithium-ion batteries. *J. Phys. Chem. C* **2014**, *118*, 21826–21833. [CrossRef]
22. Yang, F. Interaction between diffusion and chemical stresses. *Mater. Sci. Eng. A* **2005**, *409*, 153–159. [CrossRef]
23. Bower, A.F. *Applied Mechanics of Solids*; CRC Press: Boca Raton, FL, USA, 2009.
24. Ashby, M.F. Materials selection in mechanical design. *MRS Bull.* **2005**, *30*, 995. [CrossRef]
25. Renganathan, S.; Sikha, G.; Santhanagopalan, S.; White, R.E. Theoretical analysis of stresses in a lithium ion cell. *J. Electrochem. Soc.* **2010**, *157*, A155–A163. [CrossRef]
26. Satyavani, T.; Kiran, B.R.; Kumar, V.R.; Kumar, A.S.; Naidu, S.V. Effect of particle size on dc conductivity, activation energy and diffusion coefficient of lithium iron phosphate in li-ion cells. *Eng. Sci. Technol. Int. J.* **2016**, *19*, 40–44. [CrossRef]
27. Ma, Z.S.; Xie, Z.C.; Wang, Y.; Zhang, P.P.; Pan, Y.; Zhou, Y.C.; Lu, C. Failure modes of hollow core—Shell structural active materials during the lithiation—delithiation process. *J. Power Sources* **2015**, *290*, 114–122. [CrossRef]
28. Chen, M.; Sun, Q.; Li, Y.; Wu, K.; Liu, B.; Peng, P.; Wang, Q. A thermal runaway simulation on a lithium titanate battery and the battery module. *Energies* **2015**, *8*, 490–500. [CrossRef]
29. Zhu, Y.; Wang, C. Galvanostatic intermittent titration technique for phase-transformation electrodes. *J. Phys. Chem. C* **2010**, *114*, 2830–2841. [CrossRef]
30. Dash, R.; Pannala, S. Theoretical limits of energy density in silicon-carbon composite anode based lithium ion batteries. *Sci. Rep.* **2016**, *6*, 27449. [CrossRef]
31. Lithium Titanate. Available online: https://en.wikipedia.org/wiki/Lithium_titanate (accessed on 5 December 2017).
32. Julien, C.M.; Mauger, A.; Zaghib, K.; Groult, H. Comparative issues of cathode materials for li-ion batteries. *Inorganics* **2014**, *2*, 132–154. [CrossRef]

33. Kam, K.C.; Doeff, M.M. Electrode materials for lithium ion batteries. *Mater. Matters* **2012**, *7*, 56–62.
34. Chen, L.; Fan, F.; Hong, L.; Chen, J.; Ji, Y.Z.; Zhang, S.L.; Zhu, T.; Chen, L.Q. A phase-field model coupled with large elasto-plastic deformation: Application to lithiated silicon electrodes. *J. Electrochem. Soc.* **2014**, *161*, F3164–F3172. [CrossRef]
35. Lundgren, H. Thermal Aspects and Electrolyte Mass Transport in Lithium-ion Batteries. Ph.D. Thesis, KTH Royal Institute of Technology, Stockholm, Sweden, 11 June 2015.
36. Zhao, K.; Pharr, M.; Vlassak, J.J.; Suo, Z. Fracture of electrodes in lithium-ion batteries caused by fast charging. *J. Appl. Phys.* **2010**, *108*, 73517. [CrossRef]
37. Kushima, A.; Huang, J.Y.; Li, J. Quantitative fracture strength and plasticity measurements of lithiated silicon nanowires by in situ TEM tensile experiments. *ACS Nano* **2012**, *6*, 9425–9432. [CrossRef] [PubMed]
38. Qi, Y.; Hector, L.G.; James, C.; Kim, K.J. Lithium concentration dependent elastic properties of battery electrode materials from first principles calculations. *J. Electrochem. Soc.* **2014**, *161*, F3010–F3018. [CrossRef]
39. Berla, L.A.; Lee, S.W.; Cui, Y.; Nix, W.D. Mechanical behavior of electrochemically lithiated silicon. *J. Power Sources* **2015**, *273*, 41–51. [CrossRef]
40. Wolfenstine, J.; Jo, H.; Cho, Y.-H.; David, I.N.; Askeland, P.; Case, E.D.; Kim, H.; Choe, H.; Sakamoto, J. A preliminary investigation of fracture toughness of Li7La3Zr2O12 and its comparison to other solid li-ionconductors. *Mater. Lett.* **2013**, *96*, 117–120. [CrossRef]
41. Kosova, N.V.; Uvarov, N.F.; Devyatkina, E.T.; Avvakumov, E.G. Mechanochemical synthesis of LiMn$_2$O$_4$ cathode material for lithium batteries. *Solid State Ionics* **2000**, *135*, 107–114. [CrossRef]
42. Schilcher, C.; Meyer, C.; Kwade, A. Structural and electrochemical properties of calendered lithium manganese oxide cathodes. *Energy Technol.* **2016**, *4*, 1604–1610. [CrossRef]
43. Scrosati, B.; Garche, J. Lithium batteries: Status, prospects and future. *J. Power Sour.* **2010**, *195*, 2419–2430. [CrossRef]
44. Chan, C.K.; Peng, H.; Liu, G.; McIlwrath, K.; Zhang, X.F.; Huggins, R.A.; Cui, Y. High-performance lithium battery anodes using silicon nanowires. *Nat. Nanotechnol.* **2008**, *3*, 31. [CrossRef] [PubMed]
45. Rong, J.; Fang, X.; Ge, M.; Chen, H.; Xu, J.; Zhou, C. Coaxial Si/anodic titanium oxide/Si nanotube arrays for lithium-ion battery anodes. *Nano Res.* **2013**, *6*, 182–190. [CrossRef]

© 2019 by the authors. Licensee MDPI, Basel, Switzerland. This article is an open access article distributed under the terms and conditions of the Creative Commons Attribution (CC BY) license (http://creativecommons.org/licenses/by/4.0/).

Article

Submicron-Sized Nb-Doped Lithium Garnet for High Ionic Conductivity Solid Electrolyte and Performance of Quasi-Solid-State Lithium Battery

Yan Ji [1], Cankai Zhou [1], Feng Lin [1], Bingjing Li [1], Feifan Yang [1], Huali Zhu [2], Junfei Duan [1] and Zhaoyong Chen [1,*]

1. College of Materials Science and Engineering, Changsha University of Science and Technology, Changsha 410114, China; juefly@stu.csust.edu.cn (Y.J.); zhoucankai@stu.csust.edu.cn (C.Z.); 18216359528@163.com (F.L.); krystalbingjingli@163.com (B.L.); yff_0413@126.com (F.Y.); junfei_duan@csust.edu.cn (J.D.)
2. College of Physics and Electronic Science, Changsha University of Science and Technology, Changsha 410114, China; juliezhu2005@126.com
* Correspondence: chenzhaoyongcioc@126.com

Received: 23 December 2019; Accepted: 20 January 2020; Published: 24 January 2020

Abstract: The garnet $Li_7La_3Zr_2O_{12}$ (LLZO) has been widely investigated because of its high conductivity, wide electrochemical window, and chemical stability with regards to lithium metal. However, the usual preparation process of LLZO requires high-temperature sintering for a long time and a lot of mother powder to compensate for lithium evaporation. In this study submicron $Li_{6.6}La_3Zr_{1.6}Nb_{0.4}O_{12}$ (LLZNO) powder—which has a stable cubic phase and high sintering activity—was prepared using the conventional solid-state reaction and the attrition milling process, and Li stoichiometric LLZNO ceramics were obtained by sintering this powder—which is difficult to control under high sintering temperatures and when sintered for a long time—at a relatively low temperature or for a short amount of time. The particle-size distribution, phase structure, microstructure, distribution of elements, total ionic conductivity, relative density, and activation energy of the submicron LLZNO powder and the LLZNO ceramics were tested and analyzed using laser diffraction particle-size analyzer (LD), X-Ray Diffraction (XRD), Scanning Electron Microscope (SEM), Electrochemical Impedance Spectroscopy (EIS), and the Archimedean method. The total ionic conductivity of samples sintered at 1200 °C for 30 min was 5.09×10^{-4} S·cm^{-1}, the activation energy was 0.311 eV, and the relative density was 87.3%. When the samples were sintered at 1150 °C for 60 min the total ionic conductivity was 3.49×10^{-4} S·cm^{-1}, the activation energy was 0.316 eV, and the relative density was 90.4%. At the same time, quasi-solid-state batteries were assembled with $LiMn_2O_4$ as the positive electrode and submicron LLZNO powder as the solid-state electrolyte. After 50 cycles, the discharge specific capacity was 105.5 mAh/g and the columbic efficiency was above 95%.

Keywords: solid-state electrolyte; submicron powder; garnet; lithium-ion conductivity; solid-state batteries

1. Introduction

Currently, lithium-ion batteries are widely used in electric vehicles (EVs), hybrid electric vehicles (HEVs), computers, smart grids, wearable devices, etc. [1]. Traditional lithium-ion batteries have organic liquid electrolytes which easily burn and explode under abusive conditions. In addition, with the development of modern society, lithium-ion batteries have gradually moved toward high specific energy. Researchers have studied cathode materials, such as $LiCoO_2$, $LiMn_2O_4$, $LiFePO_4$, $LiNi_{0.6}Co_{0.2}Mn_{0.2}O_2$, $LiNi_{0.8}Co_{0.1}Mn_{0.1}O_2$, $LiNi_{0.8}Co_{0.15}Al_{0.05}O_2$, and $xLi_2MnO_3·(1-x)LiMO_2$, to increase the energy density of lithium-ion batteries [2–8]. The energy density can be improved by increasing the charging voltage, which will lead to serious side reactions and safety issues. In order to solve the safety problem of lithium-ion batteries, researchers have turned their attention to all-solid-state lithium batteries, which use inorganic electrolytes. The non-flammability, long cycling life and wide electrochemical window of all-solid-state lithium batteries are considered to provide the high safety and high energy density of the next-generation energy storage systems [9,10].

A solid electrolyte is an important component of all-solid-state batteries. It can not only be used as a lithium ionic conductor, substituting for a liquid organic electrolyte, but also can be used to block direct contact between the positive and negative electrodes, like a separator [11]. Solid electrolytes generally contain Li_3N, LiPON, perovskite, LISICON, NASICON, garnet, etc. [12–17]. Some of these solid electrolytes have high ionic conductivity ($\sim 10^{-3}$ S·cm^{-1}). However, some issues still exist, such as instability in an ambient atmosphere ($Li_{10}GeP_2S_{12}$, LGPS) and the metal cation being easily reduced by lithium (such as Ti^{4+} in $Li_xLa_{2/3-x/3}TiO_3$, LLTO) [18,19]. The cubic garnet LLZO was discovered by Murugan et al. [17] in 2007 and attracted world-wide attention for its advantages, e.g., the simple preparation process, high ionic conductivity ($\sim 10^{-3}$ S·cm^{-1}) at room temperature, high electrochemical window (0~6 V vs. Li/Li$^+$), and electrochemical stability of lithium metal. On the other hand, LLZO also has some defects, such as an unstable cubic phase and a low density of ceramics [20,21]. Moreover, a mass of LLZO mother powder is needed to compensate for lithium loss when sintering at high temperatures [21,22]. Many solutions have been adopted to solve the above issues. For example, Al, Ga, Fe, Ta, Nb, W, Y, and Sb doping were used to stabilize the cubic phase [23–30]; hot pressing sintering, plasma sintering, and microwave sintering were adopted to improve the relative density and sintering additives [31–33]; and Y_2O_3, Al_2O_3, B_2O_3, CaO, MgO, Li_3PO_4, and Li_4SiO_4 were investigated to reduce the grain-boundary resistance [34–40]. Usually, in order to evaluate the electrochemical performance, LLZO was used as solid electrolyte in all-solid-state batteries [41–43].

In this study, submicron LLZNO powder with a stable cubic phase was synthesized using the conventional solid-state reaction and prepared by the attrition milling process. The submicron LLZNO powder had a high sintering activity, which promoted the sintering process, reduced the sintering temperature and time, and reduced the loss of Li during high-temperature sintering. All these characteristics favored lithium stoichiometry and ionic conductivity. Furthermore, LLZNO ceramics were obtained without mother powder while sintering under reduced temperature and time. The particles-size distribution, phase structure, microtopography, total ionic conductivity, relative density, and activation energy were characterized and analyzed. The quasi-solid-state lithium batteries with $LiMn_2O_4$ as the positive electrode and submicron LLZNO powder as the solid electrolyte were assembled and the electrochemical performance are tested and analyzed.

2. Materials and Methods

2.1. The Synthesis of LLZNO Powder and Ceramics

A process flow chart of the preparation of submicron LLZNO powder and the sintering of LLZNO ceramics is showed in Figure 1. LLZNO powder was synthesized by the conventional solid-state reaction [44]. Lithium hydroxide monohydrate (LiOH·H_2O, 98%, Xilong Scientific Co., Ltd., Shantou, China), lanthanum oxide (La_2O_3, 99.99%, Shanghai Aladdin Bio-Chem Technology Co., Ltd., Shanghai, China), zirconia (ZrO_2, 99%, Shanghai Aladdin Bio-Chem Technology Co., Ltd., Shanghai, China), and niobium oxide (Nb_2O_5, 99.99%, Sinopharm Chemical Reagent Co., Ltd., Shanghai, China) were used as the raw materials and 10 wt% excess of LiOH·H_2O was added to compensate for the lithium loss in the high-temperature calcination and sintering process. Yttrium stabilized zirconia (YSZ, 4~8 mm in diameter) and isopropanol (IPA) were used as the ball-grinding medium. The ratio of raw material to grinding balls was 1:5 and the mixed raw material powder was wet-ball ground at 800 rpm in the planetary ball mill for 6 h. The mixture was dried at 70 °C for 14 h, then calcined at 950 °C for 12 h in an alumina crucible with ambient air to obtain the cubic-phase LLZNO powder. LLZNO slurry was attrition milled (Shanghai ROOT mechanical and electrical equipment Co., Ltd., Shanghai, China, 0.7 L volume, 70% filling rate) at 1000 rpm for 2 h, taking YSZ (0.4 mm in diameter) and IPA as the grinding medium, and the solid-liquid ratio was 1:5. The LLZNO slurry was dried at 70 °C for 14 h to obtain submicron LLZNO powder, from which green pellets (mass of 3 g, 19 mm in diameter and a thickness of about 4 mm) were pressed at 200 MPa under a cold uniaxial press. After that, the green pellets were sintered in a muffle furnace (Changsha Yuandong Electric Furnace Factory) without mother powder at 1100–1200 °C for 30–360 min and then cooled down naturally. At the same time, the green pellets were put on a platinum wire and placed in a crucible of MgO with the lid on to prevent impurity migration and a large amount of volatilization of lithium during the process of high-temperature sintering. For further testing, LLZNO ceramic pellets were polished with 400 and 1000 mesh sandpaper.

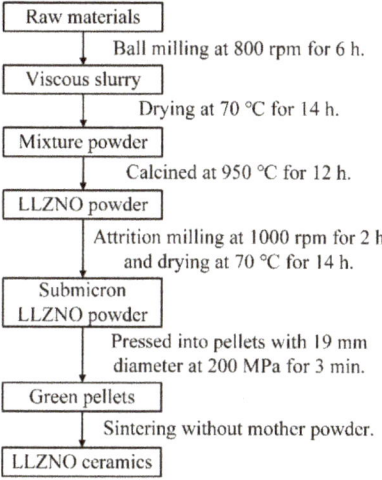

Figure 1. Process flow chart for the preparation of submicron $Li_{6.6}La_3Zr_{1.6}Nb_{0.4}O_{12}$ (LLZNO) powder and the sintering of LLZNO ceramics.

2.2. Fabrication of Composite Cathodes and Assembly of Quasi-Solid-State Batteries

In order to test the electrochemical performance of submicron LLZNO powder, we prepared a composite cathode and assembled quasi-solid-state batteries. The composite cathode consisted of a $LiMn_2O_4$ positive electrode layer and a submicron LLZNO electrolyte layer. The positive electrode was fabricated by coating the slurry of a mixture containing $LiMn_2O_4$ powder, submicron LLZNO powder, acetylene black (Shanghai Hersbit Chemical Co., Ltd., Shanghai, China), and polyvinylidene difluoride (PVDF, FR905, Shanghai San ai fu New Material Technology Co., Ltd., Shanghai, China), with a weight ratio of 7:2:1:1, onto circular aluminum foils (thickness of 20 μm, Shenzhen Kejingstar Technology Ltd., Shenzhen, China) as the current collector, and the positive material loading was 1.66 mg/cm^2. Then the composite cathode was fabricated by coating the slurry of a mixture containing submicron LLZNO powder and Polyvinylidene Fluoride (PVDF) with a weight ratio of 9:1 onto the positive electrode layer. The composite cathode was punched into disks with 18 mm diameters after compacted by a roller press (Shenzhen Kejingstar Technology Ltd., Shenzhen, China), and the density of composite cathode was about 2.5 g/cm^3. Quasi-solid-state batteries were assembled with two electrode coin cells (type CR-2025) in a glove box filled with argon and with lithium metal foil (15 mm in diameter and 1 mm thick, Shenzhen Kejingstar Technology Ltd., Shenzhen, China) as the negative current collector. In addition, 20 μL of a liquid organic electrolyte (1 M $LiPF_6$ dissolved in ethyl carbonate (EC) and dimethyl carbonate (DMC) with a ratio of 1:1, CAPCHEM, Shenzhen, China [45]) was added to improve the contact and reduce the interface impedance between the submicron LLZNO electrolyte layer and the anode/cathode [46,47]. Compared with lithium-ion batteries, the added amount of liquid organic electrolyte was small [48,49].

2.3. Characterization

X-ray diffraction (XRD, Cu-Kα radiation, λ = 1.542 Å, Bruker D8 ADVANCE, Bruker AXS GmbH, Karlsruhe, Germany) was used to determine the phase of the ceramics pellets at room temperature within 10–60 °C with steps of 5 °C/min. Jade Software was used to match and analyze the phase of the sample. The relative density of ceramics was measured by Archimedes' method and deionized water was used as the immersion medium. Meanwhile, the theoretical density of LLZNO, calculated by the Jade Software, was 5.20 g/cm^3, and the relative density was the measured density divided by the theoretical density. The particle size and distribution of the powder were determined by the laser diffraction particle-size test method (LD, Mastersizer 3000, Malvern Instruments Limited, Malvern, UK), and the relative density, refractive index, and absorption rate of the LLZNO powder was 5.20 g/cm^3, 1.4, and 0.1, respectively. The microtopography of the submicron LLZNO powder and cross section of the ceramic pellets was observed by scanning electron microscope (SEM, TESCAN MIRA3 LMU, TESCAN Orsay Holding, a. s., Brno, Czech Republic). Energy dispersive spectrometer (Oxford X-ray Max20, Oxford Instruments plc, Oxford, UK) mapping was used to characterize the distribution of each element in the cross section of the ceramic pellets. The total lithium ion conductivity of the ceramic pellets was measured by an Electrochemical Impedance Spectroscopy (EIS, Gamry Reference 600+, Gamry Instruments, Warminster, PA, USA) within a temperature range of 25–80 °C, within the frequency of 10 Hz–5 MHz, and with an AC amplitude of 40 mV. The blocking electrode was uniformly coated by a thin silver layer on both sides of the ceramic pellets. The activation energy of the ceramic pellets was measured within a temperature range of 25–80 °C and calculated based on the Arrhenius equation [16]. The quasi-solid-state batteries were tested under the battery charge-discharge tester (BTS-5V3A, Neware Technology Co., Ltd., Shenzhen, China) at 25 °C, and current density was 0.02 mA/cm^2.

3. Results and Discussions

The XRD pattern of the LLZNO powder is shown in Figure 2b and was identified as cubic phase (PDF 63-0174). The LD result and SEM image of the LLZNO powder after the attrition milling process, which demonstrated a submicron powder, are showed in Figure 2a and Table 1. The $D_{(10)}$, $D_{(50)}$, $D_{(90)}$, and primary particle size of the submicron LLZNO powder were 0.43 μm, 0.59 μm, 0.812 μm, and about 0.1 μm, respectively. The value of $D_{(3,2)}$ (0.575 μm) is similar to that of $D_{(4,3)}$ (0.607 μm), which indicates that the prepared powder had a uniform particle-size distribution. In addition, the powder also had a higher specific surface area (2007 m^2/kg), which means that the powder had a high sintering activity, which can promote crystal growth and the rapid densification of ceramics in the sintering process.

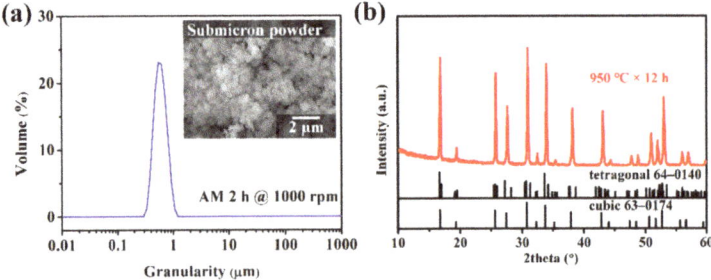

Figure 2. (**a**) Particle-size distribution of the LLZNO powder after being attrition milled 2 h at 1000 rpm and its SEM image and (**b**) XRD pattern of the LLZNO powder.

Table 1. Laser particle-size test results of submicron-scale LLZNO powder.

Preparation Condition	D_{10} (µm)	D_{50} (µm)	D_{90} (µm)	$D_{(3,2)}$ (µm)	$D_{(4,3)}$ (µm)	Specific Surface Area (m²/kg)
Attrition milled 2 h @ 1000 rpm	0.430	0.590	0.812	0.575	0.607	2007

The XRD patterns of the LLZNO ceramic samples are showed in Figure 3. The phases of all the prepared ceramic samples were identified as cubic phases (PDF 63-0174). The crystal parameters of the different samples are showed in Table 2. The XRD patterns of the samples sintered at 1200 °C × 60 min (SL1) and at 1100 °C × 360 min (SL5) showed a few impure phase peaks, mainly belonging to $LiNbO_3$ (PDF 82-0459) and Li_7NbO_6 (PDF 29-0816), and due to the decomposition from the high sintering activity of the LLZNO after having been sintered for too long at a high temperature. Moreover, these impure phases decreased the total ionic conductivity of LLZNO ceramics by increasing the resistance of the grain boundary.

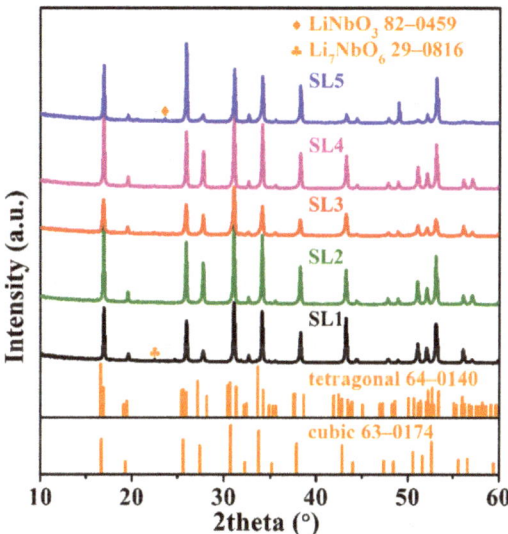

Figure 3. XRD patterns of the LLZNO ceramics with different sintering conditions.

AC impedance plots and the enlargement of the LLZNO ceramic pellets under different sintering conditions are showed in Figure 4a, b. The fitting curve of the sample sintered at 1200 °C for 30 min (SL2) is showed in Figure 4c, and it consists of a quasi-semicircle at high frequency and a long diffusion tail at low frequency. The equivalent circuit model $R_b(R_{gb}Q_{gb})(R_{el}Q_{el})$, in which R_b, R_{gb}, and R_{el} are resistances originating from the bulk, grain boundaries, and Ag electrodes, is used to fit the plots and is shown in Figure 4d. The total ionic conductivity of the ceramics is mainly decided by R_b plus R_{gb}. The total ionic conductivity and relative density of the LLZNO ceramic pellets are showed in Figure 4e and Table 2. The highest total ionic conductivity (5.09×10^{-4} S·cm^{-1}) of the LLZNO ceramic pellets was obtained when sintered at a high temperature and for a short time (SL2, 1200 °C × 30 min), and its relative density is 87.3%. This indicates that high-performance LLZNO ceramics are obtained when sintered at high temperatures only for short sintering times. However, the total ionic conductivity and relative density of ceramic pellets decreased and impure phases occurred when the sintering time was prolonged at 1100 and 1200 °C. The lowest total ionic conductivity (0.35×10^{-4} S·cm^{-1}) and relative density (83.4%) were obtained when the ceramic pellets were sintered at 1100 °C for 360 min

(SL5). Meanwhile, a higher total ionic conductivity (3.49 × 10^{-4} S·cm^{-1}) and a higher relative density (90.3%) of ceramic pellets (SL3, 1150 °C × 60 min) were obtained when sintered for 60 min from 1100 to 1200 °C, and this result indicates that, in this study, increasing the sintering temperature too much was disadvantageous for obtaining LLZNO ceramics with good performance.

Table 2. Sintering condition, cell parameter, total ionic conductivity at 25 °C, activation energy, and relative density of LLZNO ceramics.

Sample Name	Sintering Condition	Cell Parameter (Å)	Total Ionic Conductivity (10^{-4} S·cm^{-1}), 25 °C	Activation Energy (eV)	Relative Density
SL-1	1200 °C × 60 min	12.8952	1.58	0.315	86.7%
SL-2	1200 °C × 30 min	12.8953	5.09	0.311	87.3%
SL-3	1150 °C × 60 min	12.9028	3.49	0.316	90.4%
SL-4	1100 °C × 60 min	12.8916	0.51	0.319	90.3%
SL-5	1100 °C × 360 min	12.8870	0.35	0.328	83.4%

Arrhenius plots and the linear fitting curve are showed in Figure 5a. The activation energy of ceramics samples is showed in Figure 5b and Table 2, and their values are within the range of 0.31–0.33 eV. This indicates that there was no obvious effect on the activation energy of the ceramics when the green pellets prepared from the submicron LLZNO powder were sintered. The variation tendency of the activation energy was similar to the total ionic conductivity, and the lowest and the highest activation energy was 0.311 eV (SL2, 1200 °C × 30 min) and 0.328 eV (SL5, 1100 °C × 360 min).

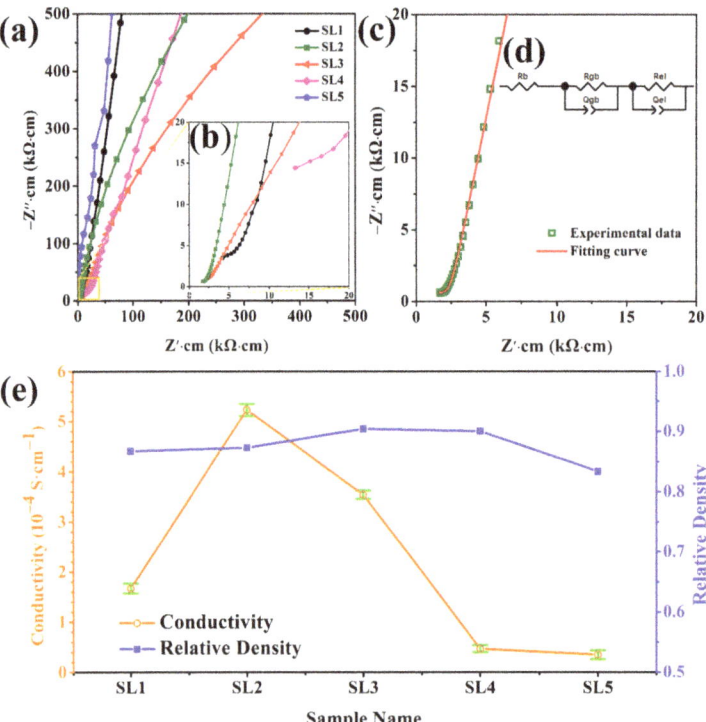

Figure 4. (**a**, **b**) AC impedance plots of the LLZNO ceramics with different sintering conditions at 25 °C; (**c**) AC impedance plots and fitting curve of SL2 at 25 °C; (**d**) equivalent circuit to fit the curves. (**e**) Total conductivity and relative density of the LLZNO ceramics.

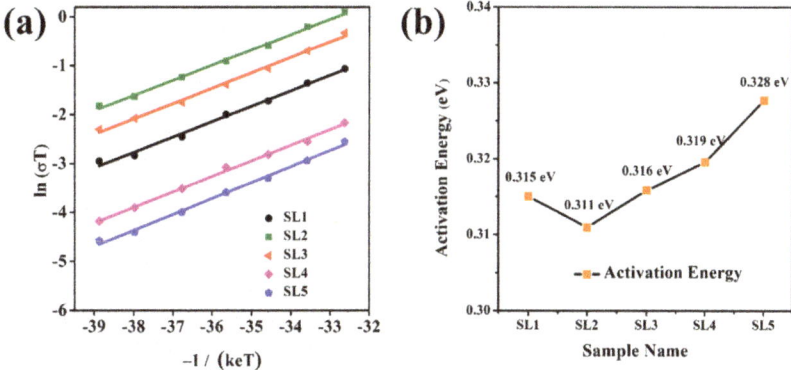

Figure 5. (a) Arrhenius plots and fitting results and (b) the activation energy of different LLZNO ceramics.

SEM images of cross sections of the LLZNO ceramics, which were sintered under different conditions, are showed in Figure 6a–e. We found that the grain size of the ceramics that were sintered for 60 min within a temperature range of 1100 to 1200 °C gradually increased from 1~5 μm (SL4, 1100 °C for 60 min, Figure 6d). A few of the grains were 5 μm and most of the grains were 100~200 μm (SL3, 1150 °C for 60 min, Figure 6c), and finally, all grains were about 200 μm (SL1, 1200 °C for 60 min, Figure 6a). Here, we found a mass of abnormal growth grains (AGGs) [50], as shown in Figure 6a,c,e, and a mass of pores were distributed in the AGGs. Meanwhile, the total ionic conductivity was lower when the AGGs were bigger. This was due to the submicron LLZNO powder having a high sintering activity, which made the crystal grain of the LLZNO ceramics have a high specific surface energy during the high-temperature sintering process, and promoted rapid grain growth and ceramic densification in the sintering process. For the above reasons, the growth rate of the grains was higher than the migration rate of the pores at the grain boundaries when the sintering temperature was higher and the sintering time was longer and the pores could not be discharged from the grain boundaries and finally stay on the inside of the AGGs. As a result, the bulk impedance of the crystal grains increased, and the total ion conductivity was reduced. However, although the submicron LLZNO powder had high sintering activity, the growth of grains could not be entirely promoted in a shorter sintering time and at a lower temperature. Therefore, a mass of grains which stayed in the initial state are shown in Figure 6d (SL4, 1100 °C × 60 min), and this was disadvantageous for lithium-ionic conduction due to the incomplete surface of the LLZNO grains after the attrition milling process. Eventually, the ceramic pellets showed a lower total ionic conductivity (0.51×10^{-4} S·cm^{-1}). A cross-sectional SEM image of the sample sintered at 1200 °C for 30 min (SL2) is showed in Figure 6b. It was found that the grains grew uniformly (~4 μm), their surfaces were smooth without pores, and they bond tightly with other grains. A highest ionic conductivity of 5.09×10^{-4} S·cm^{-1} was obtained, which indicates that the submicron LLZNO powder had a higher sintering activity and high total ionic conductivity LLZNO ceramic pellets could be obtained by sintered at a high temperature for only a short time. At the same time, the LLZNO ceramic pellets which had a higher total ionic conductivity could also be also obtained when the sintering temperature was properly reduced.

Figure 6f shows the SEM image and its EDS mapping, including La, Zr, and Nb in the cross section of the LLZNO ceramic of 1200 °C × 30 min (SL2). The cross section of the sample exhibits a transgranular fracture and an intergranular fracture, and the elements of La, Zr, and Nb are relatively uniformly distributed, which indicates that the Nb element was successfully incorporated into the LLZO lattice. This is also verified by the XRD result. However, the non-uniform distribution of Zr, La, and Nb exists in the central part of the EDS mapping. This indicates that during high-temperature sintering element segregation and depletion occurred due to the different migration rates of the elements.

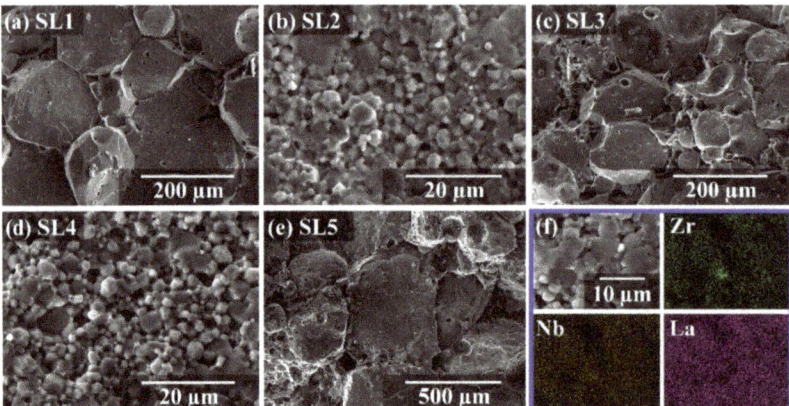

Figure 6. (a–e) SEM images of the cross-sectional microstructures of the ceramics that were sintered by different particles sizes under different sintering conditions and (f) EDS mapping of LLZNO ceramics section sintered at 1200 °C for 30 min.

The specific capacity and coulombic efficiency of quasi-solid-state batteries with $LiMn_2O_4$ as the positive electrode after 50 cycles of a galvanostatic charge-discharge test at 25 °C are showed in Figure 7a. The 1st, 2nd, 10th, 20th, and 50th galvanostatic charge-discharge curves of quasi-solid-state batteries are showed in Figure 7b. The quasi-solid-state batteries showed good cycling performance at a current density of 0.02 mA/cm^2 and a voltage within 3.0–4.3 V. The first discharge specific capacity was 106.4 mAh/g and the coulomb efficiency was 93.23%. The 2nd, 10th, 20th, and 50th discharge specific capacities were 106.8 mAh/g, 105.3 mAh/g, 106.9 mAh/g, and 105.5 mAh/g, respectively. After 50 cycles of the galvanostatic charge-discharge test, the coulomb efficiency was maintained at about 95% and the capacity retention rate was 99.15%. The capacity of the batteries increased in the early stage of the galvanostatic charge-discharge test, which may be caused by the activation of positive material. This indicates that submicron LLZNO powder can be used in quasi-solid-state batteries, and that the specific capacity and the cycling stability of quasi-solid-state batteries are relatively good. Here, the electrochemical performance of quasi-solid-state batteries using submicron LLZNO powder is only discussed, and further research will be carried out in the future.

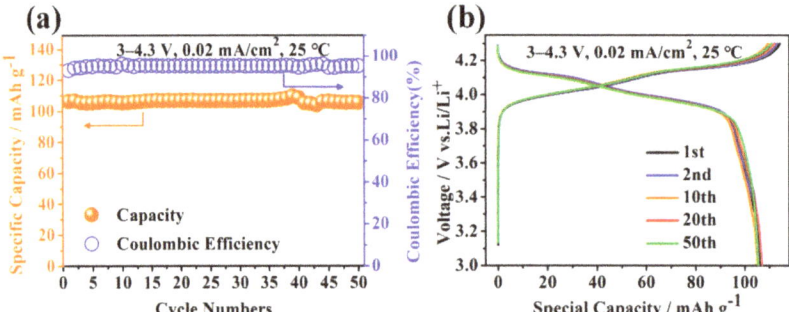

Figure 7. (a) Specific capacity and coulombic efficiency and (b) the 1st, 2nd, 10th, 20th, and 50th galvanostatic charge-discharge curves of quasi-solid-state batteries with $LiMn_2O_4$ as the positive electrode.

4. Conclusions

In this study, we synthesized Nb-doped stabilized cubic-phase LLZO powder using the conventional solid-state reaction and prepared submicron LLZNO powder using the attrition milling process. Electrolyte ceramics prepared using submicron LLZNO powder can be sintered without mother powder, which reduces the sintering temperature and shortens the sintering time. After being sintered at 1150 °C for 60 min, the total ionic conductivity, relative density, and activation energy was 3.49×10^{-4} S·cm^{-1}, 90.4%, and 0.316 eV, respectively. When sintered at 1200 °C for 30 min, we obtained the highest total ionic conductivity of 5.09×10^{-4} S·cm^{-1}, the relative density was 87.3%, and the smallest activation energy was 0.311 eV. For the quasi-solid-state batteries assembled with submicron LLZNO powder, the capacity retention rate was 99.15% and the specific capacity was 105.5 mAh/g after 50 cycles at room temperature with a current density of 0.02 mA/cm^2. Therefore, we have presented a simple method to reduce the waste of raw materials and energy used when sintering LLZO ceramics. At the same time, the prepared submicron LLZO powder can also be applied in quasi-solid-state batteries, with a good electrochemical performance.

Author Contributions: Methodology and conceptualization, Z.C., Y.J., and C.Z.; resources, Z.C. and H.Z.; data curation, Y.J., C.Z., F.L., B.L., and F.Y.; writing—original draft preparation, Y.J.; writing—review and editing, Y.J. and J.D.; funding acquisition, Z.C. and H.Z. All authors have read and agreed to the published version of the manuscript.

Funding: This work was supported by the National Natural Science Foundation of China (No. 51874048), the National Science Foundation for Young Scientists of China (No. 51604042), the Research Foundation of Education Bureau of Hunan Province (No. 19A003), the Scientific Research Fund of Changsha Science and Technology Bureau (No. kq1901100), and the Postgraduate Innovative Test Program of Hunan Province.

Conflicts of Interest: The authors declare no conflict of interest.

References

1. Zhu, C.; Wei, D.; Wu, Y.; Zhang, Z.; Zhang, G.; Duan, J.; Li, L.; Zhu, H.; Zhu, Z.; Chen, Z. Controllable construction of interconnected SnO$_x$/N-doped carbon/carbon composite for enhanced-performance lithium-ion batteries anodes. *J. Alloy. Compd.* **2019**, *778*, 731–740. [CrossRef]
2. Li, X.; Zhang, K.; Mitlin, D.; Yang, Z.; Wang, M.; Tang, Y.; Jiang, F.; Du, Y.; Zheng, J. Fundamental insight into Zr modification of Li-and Mn-rich cathodes: combined transmission electron microscopy and electrochemical impedance spectroscopy study. *Chem Mater* **2018**, *30*, 2566–2573. [CrossRef]
3. Chen, Z.; Xu, M.; Zhu, H.; Xie, T.; Wang, W.; Zhao, Q. Enhanced electrochemical performance of polyacene coated LiMn$_2$O$_{3.95}$F$_{0.05}$ for lithium ion batteries. *Appl. Surf. Sci.* **2013**, *286*, 177–183. [CrossRef]
4. Chen, Z.; Zhang, Z.; Zhao, Q.; Duan, J.; Zhu, H. Understanding the Impact of K-Doping on the Structure and Performance of LiFePO$_4$/C Cathode Materials. *J. Nanosc. Nanotechnol.* **2019**, *19*, 119–124. [CrossRef]
5. Liu, W.; Hu, G.; Du, K.; Peng, Z.; Cao, Y. Enhanced storage property of LiNi$_{0.8}$Co$_{0.15}$Al$_{0.05}$O$_2$ coated with LiCoO$_2$. *J. Power Sources* **2013**, *230*, 201–206. [CrossRef]
6. Li, X.; Zhang, K.; Wang, M.; Liu, Y.; Qu, M.; Zhao, W.; Zheng, J. Dual functions of zirconium modification on improving the electrochemical performance of Ni-rich LiNi$_{0.8}$Co$_{0.1}$Mn$_{0.1}$O$_2$. *Sustain Energ Fuels* **2018**, *2*, 413–421. [CrossRef]
7. Hu, G.; Liu, W.; Peng, Z.; Du, K.; Cao, Y. Synthesis and electrochemical properties of LiNi$_{0.8}$Co$_{0.15}$Al$_{0.05}$O$_2$ prepared from the precursor Ni$_{0.8}$Co$_{0.15}$Al$_{0.05}$OOH. *J. Power Sources* **2012**, *198*, 258–263. [CrossRef]
8. Liu, J.; Liu, Q.; Zhu, H.; Lin, F.; Ji, Y.; Li, B.; Duan, J.; Li, L.; Chen, Z. Effect of Different Composition on Voltage Attenuation of Li-Rich Cathode Material for Lithium-ion Batteries. *Materials* **2020**, *13*, 40. [CrossRef]
9. Ohta, S.; Seki, J.; Yagi, Y.; Kihira, Y.; Tani, T.; Asaoka, T. Co-sinterable lithium garnet-type oxide electrolyte with cathode for all-solid-state lithium ion battery. *J. Power Sources* **2014**, *265*, 40–44. [CrossRef]
10. Sakuda, A.; Takeuchi, T.; Kobayashi, H. Electrode morphology in all-solid-state lithium secondary batteries consisting of LiNi$_{1/3}$Co$_{1/3}$Mn$_{1/3}$O$_2$ and Li$_2$S-P$_2$S$_5$ solid electrolytes. *Solid State Ionics* **2016**, *285*, 112–117.
11. Wu, B.; Wang, S.; Evans, W.J., IV; Deng, D.Z.; Yang, J.; Xiao, J. Interfacial behaviours between lithium ion conductors and electrode materials in various battery systems. *J. Mater. Chem. A* **2016**, *4*, 15266–15280. [CrossRef]

12. Alpen, U.V.; Rabenau, A.; Talat, G. Ionic conductivity in Li$_3$N single crystals. *Appl. Phys. Lett.* **1977**, *30*, 621–623. [CrossRef]
13. Senevirathne, K.; Day, C.S.; Gross, M.D.; Lachgar, A.; Holzwarth, N. A new crystalline LiPON electrolyte: Synthesis, properties, and electronic structure. *Solid State Ionics* **2013**, *233*, 95–101. [CrossRef]
14. Uhlmann, C.; Braun, P.; Illig, J.; Weber, A.; Ivers-Tiffée, E. Interface and grain boundary resistance of a lithium lanthanum titanate (Li$_{3x}$La$_{2/3-x}$TiO$_3$, LLTO) solid electrolyte. *J. Power. Sources* **2016**, *307*, 578–586. [CrossRef]
15. Kanno, R.; Murayama, M. Lithium Ionic Conductor Thio-LISICON: The Li$_2$S·GeS$_2$·P$_2$S$_5$·System. *J. Electrochem. Soc.* **2001**, *148*, A742–A746. [CrossRef]
16. Lai, Y.; Sun, Z.; Jiang, L.; Hao, X.; Jia, M.; Wang, L.; Liu, F. Rapid sintering of ceramic solid electrolytes LiZr$_2$(PO$_4$)$_3$ and Li$_{1.2}$Ca$_{0.1}$Zr$_{1.9}$(PO$_4$)$_3$ using a microwave sintering process at low temperatures. *Ceram. Int.* **2019**, *45*, 11068–11072. [CrossRef]
17. Murugan, R.; Thangadurai, V.; Weppner, W. Fast lithium ion conduction in garnet-type Li$_7$La$_3$Zr$_2$O$_{12}$. *Angew. Chem. Int. Edit.* **2007**, *46*, 7778–7781. [CrossRef]
18. Famprikis, T.; Canepa, P.; Dawson, J.A.; Islam, M.S.; Masquelier, C. Fundamentals of inorganic solid-state electrolytes for batteries. *Nat. Mater.* **2019**, *18*, 1278–1291. [CrossRef]
19. Inaguma, Y.; Nakashima, M. A rechargeable lithium–air battery using a lithium ion-conducting lanthanum lithium titanate ceramics as an electrolyte separator. *J. Power Sources* **2013**, *228*, 250–255. [CrossRef]
20. Awaka, J.; Kijima, N.; Kataoka, K.; Hayakawa, H.; Ohshima, K.-i.; Akimoto, J. Neutron powder diffraction study of tetragonal Li$_7$La$_3$Hf$_2$O$_{12}$ with the garnet-related type structure. *J. Solid State Chem.* **2010**, *183*, 180–185. [CrossRef]
21. Huang, Z.; Liu, K.; Chen, L.; Lu, Y.; Li, Y.; Wang, C.A. Sintering behavior of garnet-type Li$_{6.4}$La$_3$Zr$_{1.4}$Ta$_{0.6}$O$_{12}$ in Li$_2$CO$_3$ atmosphere and its electrochemical property. *Int. J. Appl. Ceram. Technol.* **2017**, *14*, 921–927. [CrossRef]
22. Ren, Y.; Deng, H.; Chen, R.; Shen, Y.; Lin, Y.; Nan, C.-W. Effects of Li source on microstructure and ionic conductivity of Al-contained Li$_{6.75}$La$_3$Zr$_{1.75}$Ta$_{0.25}$O$_{12}$ ceramics. *J. Eur. Ceram. Soc.* **2015**, *35*, 561–572. [CrossRef]
23. Tsai, C.-L.; Dashjav, E.; Hammer, E.-M.; Finsterbusch, M.; Tietz, F.; Uhlenbruck, S.; Buchkremer, H.P. High conductivity of mixed phase Al-substituted Li$_7$La$_3$Zr$_2$O$_{12}$. *J. Electroceram.* **2015**, *35*, 25–32. [CrossRef]
24. Janani, N.; Deviannapoorani, C.; Dhivya, L.; Murugan, R. Influence of sintering additives on densification and Li+ conductivity of Al doped Li$_7$La$_3$Zr$_2$O$_{12}$ lithium garnet. *RSC Adv.* **2014**, *4*, 51228–51238. [CrossRef]
25. Jin, Y.; McGinn, P.J. Al-doped Li$_7$La$_3$Zr$_2$O$_{12}$ synthesized by a polymerized complex method. *J. Power Sources* **2011**, *196*, 8683–8687. [CrossRef]
26. Huang, M.; Shoji, M.; Shen, Y.; Nan, C.-W.; Munakata, H.; Kanamura, K. Preparation and electrochemical properties of Zr-site substituted Li$_7$La$_3$(Zr$_{2-x}$M$_x$)O$_{12}$ (M = Ta, Nb) solid electrolytes. *J. Power Sources* **2014**, *261*, 206–211. [CrossRef]
27. Ohta, S.; Kobayashi, T.; Asaoka, T. High lithium ionic conductivity in the garnet-type oxide Li$_{7-X}$La$_3$(Zr$_{2-X}$, Nb$_X$)O$_{12}$ (X = 0–2). *J. Power Sources* **2011**, *196*, 3342–3345. [CrossRef]
28. Cao, Z.-Z.; Ren, W.; Liu, J.-R.; Li, G.-R.; Gao, Y.-F.; Fang, M.-H.; He, W.-Y. Microstructure and ionic conductivity of Sb-doped Li$_7$La$_3$Zr$_2$O$_{12}$ ceramics. *J. Inorg. Mater.* **2014**, *29*, 220–224. [CrossRef]
29. Deviannapoorani, C.; Shankar, L.S.; Ramakumar, S.; Murugan, R. Investigation on lithium ion conductivity and structural stability of yttrium-substituted Li$_7$La$_3$Zr$_2$O$_{12}$. *Ionics* **2016**, *22*, 1281–1289. [CrossRef]
30. Mukhopadhyay, S.; Thompson, T.; Sakamoto, J.; Huq, A.; Wolfenstine, J.; Allen, J.L.; Bernstein, N.; Stewart, D.A.; Johannes, M. Structure and stoichiometry in supervalent doped Li$_7$La$_3$Zr$_2$O$_{12}$. *Chem. Mater.* **2015**, *27*, 3658–3665. [CrossRef]
31. David, I.N.; Thompson, T.; Wolfenstine, J.; Allen, J.L.; Sakamoto, J. Microstructure and Li-Ion Conductivity of Hot-Pressed Cubic Li$_7$La$_3$Zr$_2$O$_{12}$. *J. Am. Ceram. Soc.* **2015**, *98*, 1209–1214. [CrossRef]
32. Baek, S.W.; Lee, J.M.; Kim, T.Y.; Song, M.S.; Park, Y. Garnet related lithium ion conductor processed by spark plasma sintering for all solid state batteries. *J. Power Sources* **2014**, *249*, 197–206. [CrossRef]
33. Amores, M.; Ashton, T.E.; Baker, P.J.; Cussen, E.J.; Corr, S.A. Fast microwave-assisted synthesis of Li-stuffed garnets and insights into Li diffusion from muon spin spectroscopy. *J. Mater. Chem. A* **2016**, *4*, 1729–1736. [CrossRef]
34. Murugan, R.; Ramakumar, S.; Janani, N. High conductive yttrium doped Li$_7$La$_3$Zr$_2$O$_{12}$ cubic lithium garnet. *Electrochem. Commun.* **2011**, *13*, 1373–1375. [CrossRef]

35. Kumazaki, S.; Iriyama, Y.; Kim, K.-H.; Murugan, R.; Tanabe, K.; Yamamoto, K.; Hirayama, T.; Ogumi, Z. High lithium ion conductive $Li_7La_3Zr_2O_{12}$ by inclusion of both Al and Si. *Electrochem. Commun.* **2011**, *13*, 509–512. [CrossRef]
36. Li, Y.; Cao, Y.; Guo, X. Influence of lithium oxide additives on densification and ionic conductivity of garnet-type $Li_{6.75}La_3Zr_{1.75}Ta_{0.25}O_{12}$ solid electrolytes. *Solid State Ionics* **2013**, *253*, 76–80. [CrossRef]
37. Tadanaga, K.; Takano, R.; Ichinose, T.; Mori, S.; Hayashi, A.; Tatsumisago, M. Low temperature synthesis of highly ion conductive $Li_7La_3Zr_2O_{12}$–Li_3BO_3 composites. *Electrochem. Commun.* **2013**, *33*, 51–54. [CrossRef]
38. Janani, N.; Ramakumar, S.; Kannan, S.; Murugan, R. Optimization of lithium content and sintering aid for maximized Li^+ conductivity and density in Ta-doped $Li_7La_3Zr_2O_{12}$. *J. Am. Ceram. Soc.* **2015**, *98*, 2039–2046. [CrossRef]
39. Jonson, R.A.; McGinn, P.J. Tape casting and sintering of $Li_7La_3Zr_{1.75}Nb_{0.25}Al_{0.1}O_{12}$ with Li_3BO_3 additions. *Solid State Ionics* **2018**, *323*, 49–55.
40. Rosero-Navarro, N.C.; Yamashita, T.; Miura, A.; Higuchi, M.; Tadanaga, K. Effect of sintering additives on relative density and Li-ion conductivity of Nb-doped $Li_7La_3Zr_2O_{12}$ solid electrolyte. *J. Am. Ceram. Soc.* **2017**, *100*, 276–285. [CrossRef]
41. Wu, J.-F.; Pang, W.K.; Peterson, V.K.; Wei, L.; Guo, X. Garnet-type fast Li-ion conductors with high ionic conductivities for all-solid-state batteries. *ACS Appl. Mater. Interfaces* **2017**, *9*, 12461–12468. [CrossRef] [PubMed]
42. Ohta, S.; Kobayashi, T.; Seki, J.; Asaoka, T. Electrochemical performance of an all-solid-state lithium ion battery with garnet-type oxide electrolyte. *J. Power Sources* **2012**, *202*, 332–335. [CrossRef]
43. Jin, Y.; McGinn, P.J. $Li_7La_3Zr_2O_{12}$ electrolyte stability in air and fabrication of a Li/ $Li_7La_3Zr_2O_{12}$/$Cu_{0.1}V_2O_5$ solid-state battery. *J. Power Sources* **2013**, *239*, 326–331. [CrossRef]
44. Hu, Z.; Liu, H.; Ruan, H.; Hu, R.; Su, Y.; Zhang, L. High Li-ion conductivity of Al-doped $Li_7La_3Zr_2O_{12}$ synthesized by solid-state reaction. *Ceram. Int.* **2016**, *42*, 12156–12160. [CrossRef]
45. Liu, Q.; Zhu, H.; Liu, J.; Liao, X.; Tang, Z.; Zhou, C.; Yuan, M.; Duan, J.; Li, L.; Chen, Z. High-Performance Lithium-Rich Layered Oxide Material: Effects of Preparation Methods on Microstructure and Electrochemical Properties. *Materials* **2020**, *13*, 334. [CrossRef]
46. Zhang, W.; Nie, J.; Li, F.; Wang, Z.L.; Sun, C. A durable and safe solid-state lithium battery with a hybrid electrolyte membrane. *Nano Energy* **2018**, *45*, 413–419. [CrossRef]
47. Yu, J.; Kwok, S.C.; Lu, Z.; Effat, M.B.; Lyu, Y.Q.; Yuen, M.M.; Ciucci, F. A Ceramic-PVDF Composite Membrane with Modified Interfaces as an Ion-Conducting Electrolyte for Solid-State Lithium-Ion Batteries Operating at Room Temperature. *Chem. Electro. Chem.* **2018**, *5*, 2873–2881. [CrossRef]
48. Ye, M.; Jin, X.; Nan, X.; Gao, J.; Qu, L. Paraffin wax protecting 3D non-dendritic lithium for backside-plated lithium metal anode. *Energy Storage Mater.* **2020**, *24*, 153–159. [CrossRef]
49. Manzi, J.; Brutti, S. Surface chemistry on $LiCoPO_4$ electrodes in lithium cells: SEI formation and self-discharge. *Electrochim. Acta* **2016**, *222*, 1839–1846. [CrossRef]
50. Huang, X.; Xiu, T.; Badding, M.E.; Wen, Z. Two-step sintering strategy to prepare dense Li-Garnet electrolyte ceramics with high Li^+ conductivity. *Ceram. Int.* **2018**, *44*, 5660–5667. [CrossRef]

© 2020 by the authors. Licensee MDPI, Basel, Switzerland. This article is an open access article distributed under the terms and conditions of the Creative Commons Attribution (CC BY) license (http://creativecommons.org/licenses/by/4.0/).

Article

Structure and Electrochemical Properties of Mn₃O₄ Nanocrystal-Coated Porous Carbon Microfiber Derived from Cotton

Dongya Sun [1,2,*], Liwen He [1,2], Yongle Lai [1], Jiqiong Lian [1], Jingjing Sun [1], An Xie [1] and Bizhou Lin [2,*]

[1] Key Laboratory of Functional Materials and Applications of Fujian Province, School of Materials Science and Engineering, Xiamen University of Technology, Xiamen 361024, China; heliwen@hqu.edu.cn (L.H.); 2013123201@xmut.edu.cn (Y.L.); 2013123202@xmut.edu.cn (J.L.); 2015000093@xmut.edu.cn (J.S.); anxie@xmut.edu.cn (A.X.)
[2] Fujian Key Laboratory of Photoelectric Functional Materials, College of Materials Science and Engineering, Huaqiao University, Xiamen 361021, China
* Correspondence: 2013123205@xmut.edu.cn (D.S.); bzlinhqu@126.com (B.L.)

Received: 16 September 2018; Accepted: 12 October 2018; Published: 15 October 2018

Abstract: Biomorphic Mn₃O₄ nanocrystal/porous carbon microfiber composites were hydrothermally fabricated and subsequently calcined using cotton as a biotemplate. The as-prepared material exhibited a specific capacitance of 140.8 $F·g^{-1}$ at 0.25 $A·g^{-1}$ and an excellent cycle stability with a capacitance retention of 90.34% after 5000 cycles at 1 $A·g^{-1}$. These characteristics were attributed to the introduction of carbon fiber, the high specific surface area, and the optimized microstructure inherited from the biomaterial.

Keywords: Mn₃O₄; carbon microfibers; biotemplate; microstructure; energy storage and conversion; electrochemical properties

1. Introduction

Electrochemical supercapacitors (ESs) have many desirable properties, including long lifetime, high power density, and high rate capability. Thus, ESs are attracting worldwide attention as an efficient energy storage device for portable electronic devices and vehicles [1,2]. Transition metal oxides (TMOs) such as CoO_x, NiO, and MnO_x have been extensively investigated as promising electrode materials for ES applications. These oxides deliver higher specific capacitances than those of carbonaceous materials due to reversible faradaic redox reactions [2]. Manganese oxide has been considered a highly attractive TMO due to its high theoretical specific capacitance, good electrochemical stability, low cost, and natural abundance [3]. MnO_x-based composites with various microstructures and morphologies, such as wires, sheets, tubes, and flowers, have been developed [4–7]. Although these active materials exhibit enhanced pseudocapacitance properties, their low electronic conductivity and insufficient interface contact can substantially reduce the experimental specific capacitance and hamper their extensive commercial application.

To address these issues, scholars have proposed the fabrication of many MnO_x/carbon composites in recent years [8,9]. Among the carbon materials, using low-cost natural resources, biowaste, and food waste are highly effective ways to achieve the large-scale production of electrode materials [10–12]. Natural biomaterials usually possess irregular microstructures that are difficult to duplicate, and these structures are often highly suitable as energy storage materials. For example, cotton wool [6], coconut shell [13], and human hair [14] have been introduced as excellent flexible carbon substrates for hybrid composites. These materials exhibit high specific capacitance, good electronic conductivity, good cycle stability, and excellent flexibility. Therefore, a simple and cheap approach using porous carbon microfiber (PCM) derived from cotton wool was studied in our work.

Herein, we prepared a biomorphic porous composite fabricated using Mn_3O_4 nanocrystals supported on PCMs (Mn_3O_4/PCM) with cotton fiber as the biotemplate and carbon skeleton. Benefitting from the unique morphology, large specific surface area, and existence of carbon fiber, the as-prepared composites exhibit an excellent electrochemical performance.

2. Experimental

The preparation process of Mn_3O_4/PCM is schematically illustrated in Scheme 1. Dry cotton wool (Xinghua health cotton wool Co. LTD, Xinghua, China) was cut into short pieces and carbonized for 1 h at 350 °C under an argon atmosphere. The as-prepared sample is denoted as PCM. Then, 2.5 g PCM was immersed in 50 mL of 1 mol·L^{-1} $Mn(NO_3)_2$·$6H_2O$ solution. The dispersed solution was transferred to a 100 mL autoclave and reacted for 12 h at 180 °C. The hydrothermal reaction product was filtered, washed, dried, and annealed in a tubular furnace at 400 °C in air for 1 h, and the product was labeled as Mn_3O_4/PCM. For comparison, pristine Mn_3O_4 was hydrothermally synthesized by decomposing the solution of $Mn(NO_3)_2$·$6H_2O$ at 180 °C for 12 h without cotton.

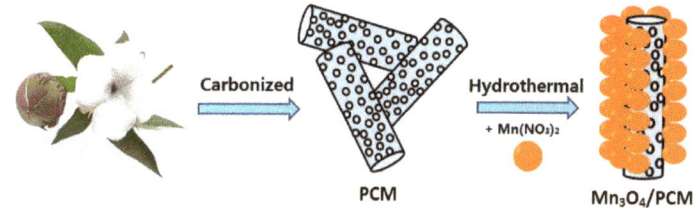

Scheme 1. Illustration of the preparation process of Mn_3O_4/PCM (porous carbon microfiber).

Field emission scanning electron microscopy (FE-SEM) images were taken with a Zeiss Sigma 500 microscope (Carl Zeiss A G, Jena, German) and an Oxford X-Max energy dispersive spectroscope (EDS, Oxford instruments, Oxford, UK). Transmission electron microscopy (TEM) images were taken using a FEI Talos F200S microscope (Thermo fisher Scientific, Waltham, MA, USA) with an accelerating voltage of 200 kV. Powder X-ray diffraction (XRD) patterns were obtained at ambient temperature on a Rigaku Smart Lab 3 kW diffractometer (Rigaku, Tokyo, Japan) using Cu K radiation (λ = 1.5418 Å) under an accelerating voltage of 36 kV. Raman spectroscopy was tested with the range of 300–2000 cm^{-1} on a LABRAM HR-800 spectrometer (Horiba, Kyoto, Japan), and the excitation source was a 532 nm laser. Specific surface area and porosity measurements were carried out on a Quantachrome Autosorb-iQ instrument (Quantachrome instruments, Boynton Beach, FL, USA) using the Brunauer–Emmett–Teller (BET) method. X-ray photoelectron spectroscopy (XPS) measurements were performed on a VG Escalab MK II spectrometer (Thermo fisher Scientific, Waltham, MA, USA) with non-monochromatic Al Kα X-ray (1486.6 eV). The TG-DTA were measured on an integrated thermal analyzer (TG, STA 449C, NETZSCH-Gerätebau GmbH, Selb, Germany) with a 10 °C/min heating rate from room temperature to 750 °C under air atmosphere. The working electrode was prepared by mixing the prepared composites, acetylene black, and polytetrafluoroethylene with a mass ratio of 75:15:10, while electrode of the bulk Mn_3O_4 was 50:40:10. The net weight of Mn_3O_4 in electrodes was ~3 mg. Subsequently, the mixture was coated onto a nickel foam current collector (1.5 cm^2), pressed at 10 MPa, and dried under vacuum at 60 °C. All the measurements were performed with a three-electrode system (Hg/HgO and platinum as the reference and counter electrode, respectively) and a two-electrode system (Mn_3O_4/PCM and commercial active carbon (CAC)) in 3 mol·L^{-1} KOH aqueous electrolyte.

3. Results and Discussion

The carbon fiber of the cotton became remarkably porous after carbonization (Figure 1a). The Mn$_3$O$_4$/PCM composite prepared using cotton as template had a tubular morphology (Figure 1b,c) similar to cotton. At the start of the hydrothermal reaction, the cotton was immersed into the precursor solution, the Mn^{2+} was partly oxidized to Mn^{3+} ions, and Mn$_3$O$_4$ formed. Mn ions were then bonded onto the oxygen-rich group of the cotton fibers through O-H, C=O groups, and so on (Figure 1b) [15–17]. The size of nanocrystals increased from several to dozens of nanometers (Figure 1b,c) with a reaction time from 5 h to 12 h. When the hydrothermal product was calcined at 400 °C, the precise replication of cotton texture could be achieved after the removal of most of the template substance. Part of the organic template was carbonized into a porous carbon stick with irregular pores (Figure 1d). The high-resolution TEM (HRTEM) image (Figure 1e) clearly demonstrates that the biomorphic Mn$_3$O$_4$ was monocrystalline and derived from the lattice arrangement (inset of Figure 1e). The interplanar spacing was about 0.25 nm for the {211} lattice planes. The distribution of Mn (44.8 wt.%), C (31.7 wt.%), and O (23.5 wt.%) in the as-prepared sample was uniform (Figure 1f).

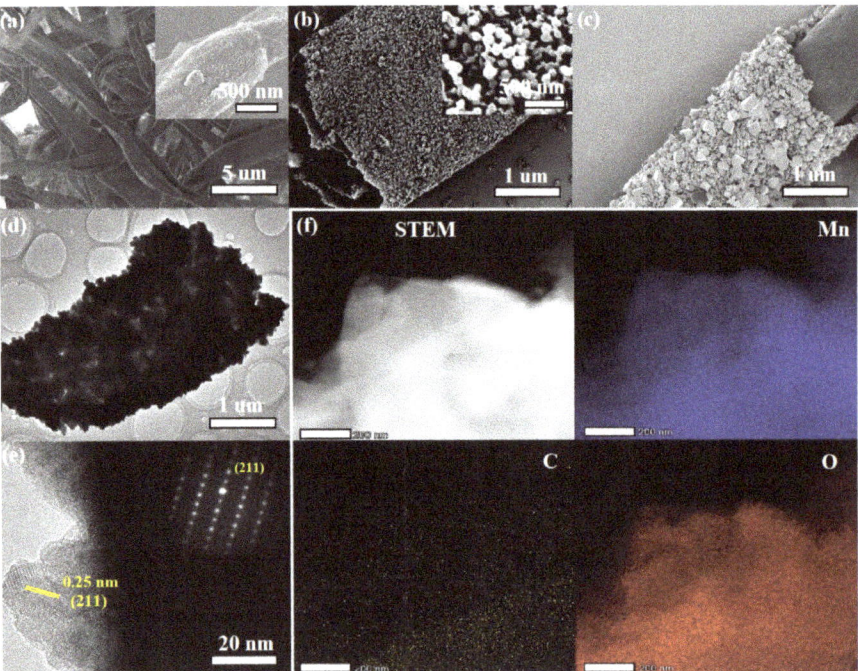

Figure 1. SEM images of (**a**) cotton and carbonized cotton fiber (inset in a); (**b**,**c**) Mn$_3$O$_4$ nanocrystals grown on the PCM surface after 5 and 12 h, respectively. (**d**) Transmission electron microscopy (TEM) image, (**e**) high-resolution TEM (HRTEM) image (inset is the SEAD graph), and (**f**) energy dispersive spectroscopy (EDS) mappings of Mn$_3$O$_4$/PCM.

The XRD patterns of Mn$_3$O$_4$/PCM and bulk Mn$_3$O$_4$ are shown in Figure 2a. The diffraction peaks of the composite can be ascribed to tetragonal Mn$_3$O$_4$ (JCPDS 18-0803) [18], and the diffraction peak at 25.24° was ascribed to the carbon fabric (graphitic carbon: JCPDS75-1621), which was marked with an asterisk symbol (*). Raman characterizations (Figure 2b) showed features from Mn–O in 600–700 cm^{-1} and carbon fibers in the 1200-1600 cm^{-1} region in Mn$_3$O$_4$/PCM [4,9,16], while only an Mn–O peak at lower wavenumber in Mn$_3$O$_4$.

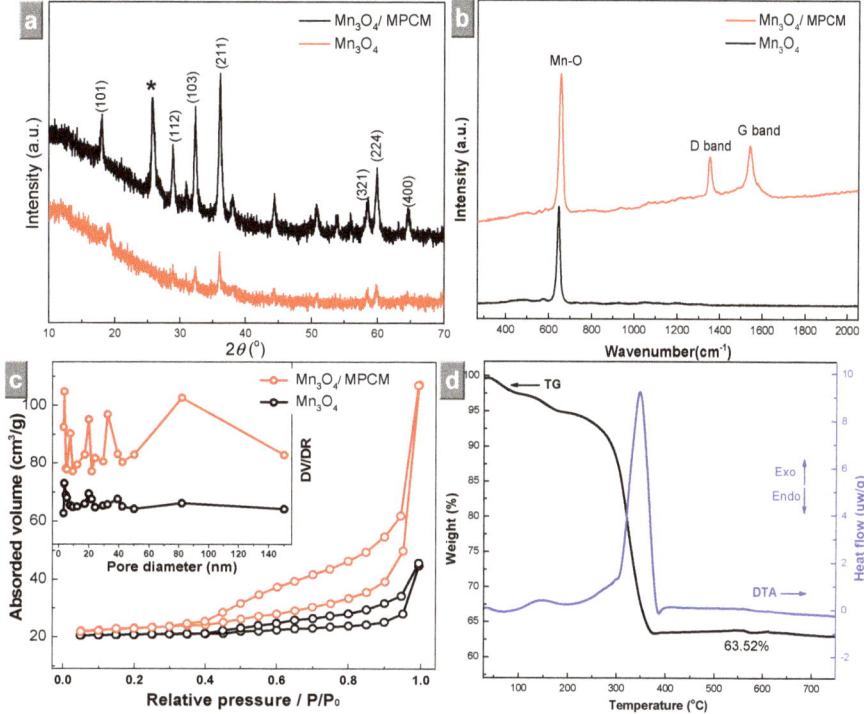

Figure 2. (**a**) XRD patterns, (**b**) Raman spectra, (**c**) nitrogen adsorption–desorption isotherms (inset shows pore size distributions) of Mn$_3$O$_4$/PCM and bulk Mn$_3$O$_4$, and (**d**) TG-DTA curves of the product.

The N$_2$ adsorption–desorption isotherm of Mn$_3$O$_4$/PCM exhibited a distinct H3 hysteresis loop at $p \cdot p_0^{-1} > 0.40$ (Figure 2c). The isotherms of composite and pristine Mn$_3$O$_4$ exhibited features of mesoporous materials. The pore size distribution of Mn$_3$O$_4$/PCM appeared with a wide range having two pore extremes at 24.2 and 82.5 nm. The BET specific surface area of the Mn$_3$O$_4$/PCM was determined to be 51.9 m$^2 \cdot$g^{-1}, which was much larger than that of pristine Mn$_3$O$_4$ (10.8 m$^2 \cdot$g^{-1}). The higher surface area and pore volume is beneficial to the contact of the electrolytes with the active materials, and can further increase the electrochemical properties [6,8,10]. The TG-DTA curves of the hydrothermal product are shown in Figure 2d. Two endothermic peaks were observed in the range of 100-200 °C because of the adsorbed water. A weight loss of ~30% was found between 250 and 380 °C, which can be attributed to the decomposition of organic substances in cotton similar to other biomorphic oxides [13,14].

As shown in Figure 3a,b, the cyclic voltammetry (CV) and galvanostatic charge–discharge (GCD) curves of Mn$_3$O$_4$/PCM and bare Mn$_3$O$_4$ demonstrated that the electrochemical properties of the former were better than those of the latter. The specific capacitances were calculated to be 120.8 and 52.2 F·g^{-1}, respectively, which were attributed to the higher specific surface area of Mn$_3$O$_4$/PCM, the existence of carbon fiber, and its special hierarchical porous morphology derived from the biotemplate. As illustrated in Figure 3c, Mn$_3$O$_4$/PCM and pristine Mn$_3$O$_4$ had a pair of redox peaks located at ~0.25 and ~0.4 V, which correspond to almost the same component and chemical conversion between different manganese oxidation states in alkaline medium (Mn$_3$O$_4$ ⇔ MnOOH) [18,19]. The GCD curves of Mn$_3$O$_4$/PCM displayed remarkable pseudocapacitance properties. A slight reduction of specific capacitance was noted at high current densities (Figure 3d), indicating that the as-prepared composite has a good electrochemical stability [4,14].

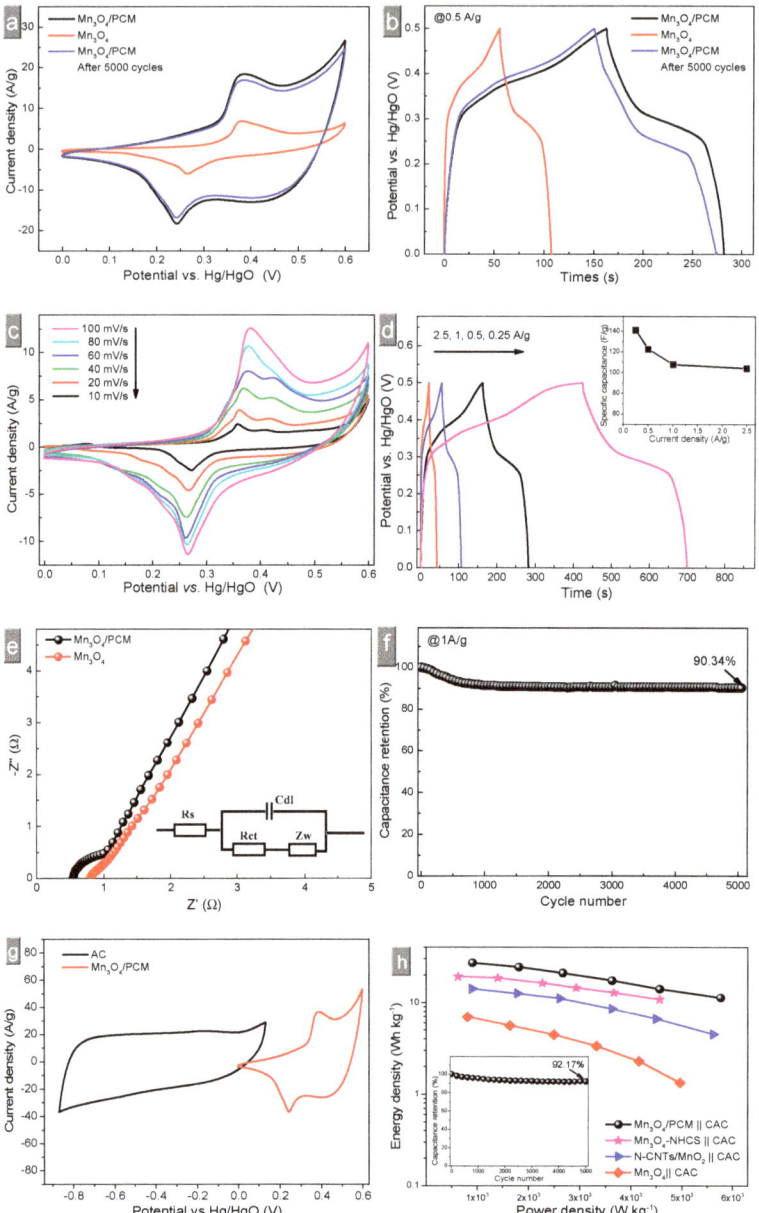

Figure 3. (**a**) Cyclic voltammetry (CV) curves at a scan rate of 80 mV·s^{-1}; (**b**) galvanostatic charge–discharge (GCD) curves at 0.5 A·g^{-1} of Mn$_3$O$_4$/PCM and bare Mn$_3$O$_4$; (**c**) CV curves of Mn$_3$O$_4$/PCM at various scan rates; (**d**) GCD curves and specific capacitance at various current densities of Mn$_3$O$_4$/PCM (inset in d); (**e**) Nyquist plots of Mn$_3$O$_4$/PCM and bare Mn$_3$O$_4$; (**f**) capacitance retention of Mn$_3$O$_4$/PCM vs. cycle number at 1 A·g^{-1}; (**g**) CV curves of Mn$_3$O$_4$/PCM and CAC at 20 mV·s^{-1}; and (**h**) Ragone plots of asymmetrical supercapacitor in MnO$_x$-based two-electrodes systems. CAC: commercial active carbon.

The electrochemical impedance spectroscopy (EIS) measurements of the biomorphic Mn_3O_4/PCM and bulk Mn_3O_4 are shown in Figure 3e. The charge-transfer resistance (R_{ct}) (radius of the semicircle) of biomorphic electrode ($\leq 0.5\ \Omega$) was obviously lower than that of pristine Mn_3O_4 (>0.6 Ω), suggesting the substantially improved reaction kinetics (e.g., increased charge transfer rate) for microtubular Mn_3O_4/PCM. The increase in rate enables additional rapid redox reaction and facilitates electron transport, and thus improves the specific capacitance. As shown in Figure 3f, the long-term cycling stability of Mn_3O_4/PCM electrode was tested at 1 $A \cdot g^{-1}$ for 5000 cycles. This result indicates that the electrode maintained more than 90% of its initial value after 5000 cycles, suggesting the excellent cyclic stability of the electrode. At the same time, the CV and GCD profiles after 5000 cycles migrated slightly from values before cycling (Figure 3a,b). The superior electrochemical performance of Mn_3O_4/PCM can be attributed to the special tubular morphology, high specific surface area, and porous carbon microfibers inside.

After assembling into an asymmetric supercapacitor using the two-electrode system of Mn_3O_4/PCM and CAC, the working voltage could be expanded to 0–1.4 V (Figure 3g) while the three-electrode system was 0–0.6 V. As shown in Figure 3h, the specific capacitance of Mn_3O_4/PCM||CAC was 110.8 $F \cdot g^{-1}$ at a current density of 1 $A \cdot g^{-1}$. The inset of Figure 3h is a Ragone plot of the asymmetric electrode. The energy density of the cell configuration was 27.13 $Wh \cdot kg^{-1}$ at a power density of 0.41 $kW \cdot kg^{-1}$. Even at a high power density of 4.76 $kW \cdot kg^{-1}$, the energy density still maintained at 11.34 $Wh \cdot kg^{-1}$. Our results were better than previous reports concerned with special morphologies of Mn_3O_4-based composites used as supercapacity electrodes (Figure 3h) [4,20]. There was still specific capacitance of 92.17% retained in the cell after 5000 cycles (inset of Figure 3h), suggesting that the electrode has a high reversibility and good electrochemical stability.

4. Conclusions

This work successfully prepared a biomorphic Mn_3O_4/PCM composite via a simple hydrothermal route by using cotton as a biotemplate. The as-obtained Mn_3O_4/PCM was composed of Mn_3O_4 nanocrystal-coated carbon fiber, inherited the morphology and microstructure of cotton, and exhibited the excellent electrochemical properties of a specific capacitance of 140.8 $F \cdot g^{-1}$ at 0.25 $A \cdot g^{-1}$ and long cycling life with 90.34% of the capacitance after 5000 cycles. This work provides an example of the fabrication of biomorphic porous TMOs for developing promising electrode materials in energy storage and conversion.

Author Contributions: Original draft preparation, D.S.; review and editing, B.L.; literature search, L.H.; data collection, Y.L., J.L. and J.S.; data analysis, A.X.

Funding: This research was funded by the Natural Science Foundation of Fujian Province of China, grant number 2014H0028, Educational research projects for young and middle-aged teachers in Fujian Province, grant number JT180422, Program for the Innovative Research Team in Science and Technology in Fujian Province University, grant number 2018 and the Open Fund of open fund of Fujian Provincial Key Laboratory of Functional Materials and Applications (Xiamen University of Technology), grant number 608160030215.

Conflicts of Interest: The authors declare no conflict of interest. The funders had no role in the design of the study; in the collection, analyses, or interpretation of data; in the writing of the manuscript, or in the decision to publish the results.

References

1. Zhang, Y.Z.; Wang, Y.; Cheng, T.; Lai, W.Y.; Pang, H.; Huang, W. Flexible supercapacitors based on paper substrates: A new paradigm for low-cost energy storage. *Chem. Soc. Rev.* **2015**, *44*, 5181–5199. [CrossRef] [PubMed]
2. Li, B.; Dai, F.; Xiao, Q.; Yang, L.; Shen, J.; Zhang, C.; Cai, M. Nitrogen-doped activated carbon for high energy hybrid supercapacitor. *Energy Environ. Sci.* **2016**, *9*, 102–106. [CrossRef]
3. Huang, M.; Li, F.; Dong, F.; Zhang, Y.X.; Zhang, L.L. MnO_2-based nanostructures for high-performance supercapacitors. *J. Mater. Chem. A* **2015**, *3*, 21380–21423. [CrossRef]

4. Liu, T.; Jiang, C.; You, W.; Yu, J. Hierarchical porous C/MnO$_2$ composite hollow microspheres with enhanced supercapacitor performance. *J. Mater. Chem. A* **2017**, *5*, 8635–8643. [CrossRef]
5. Zang, J.; Ye, J.C.; Qian, H.; Lin, Y.; Zhang, X.W.; Zheng, M.S.; Dong, Q.F. Hollow carbon sphere with open pore encapsulated MnO$_2$ nanosheets as high-performance anode materials for lithium ion batteries. *Electrochim. Acta* **2018**, *260*, 783–788. [CrossRef]
6. Yan, D.L.; Li, S.C.; Zhu, G.S.; Wang, Z.M.; Xu, H.R.; Yu, A.B. Synthesis and pseudocapacitive behaviors of biomorphic mesoporous tubular MnO$_2$ templated from cotton. *Mater. Lett.* **2013**, *95*, 164–167. [CrossRef]
7. Zhou, Y.; Guo, L.; Shi, W.; Zou, X.F.; Xiang, B.; Xing, S.H. Rapid production of Mn$_3$O$_4$/rGO as an efficient electrode material for supercapacitor by flame plasma. *Materials* **2018**, *11*, 881. [CrossRef] [PubMed]
8. Aveiro, L.R.; Silva, A.G.M.D.; Antonin, V.S.; Candido, E.G.; Parreira, L.S.; Geonmonond, R.S.; Freitas, I.C.D.; Lanza, M.R.V.; Camargo, P.H.C.; Santos, M.C. Carbon-supported MnO$_2$ nanoflowers: Introducing oxygen vacancies for optimized volcano-type electrocatalytic activities towards H$_2$O$_2$ generation. *Electrochim. Acta* **2018**, *268*, 101–110. [CrossRef]
9. Yan, D.L.; Zhang, H.; Chen, L.; Zhu, G.S.; Li, S.C.; Xu, H.R.; Yu, A.B. Biomorphic synthesis of mesoporous Co$_3$O$_4$ microtubules and their pseudocapacitive performance. *ACS Appl. Mater. Interfaces* **2014**, *6*, 15632–15637. [CrossRef] [PubMed]
10. Wang, J.; Nie, P.; Ding, B.; Dong, S.; Hao, X.; Dou, H.; Zhang, X. Biomass derived carbon for energy storage devices. *J. Mater. Chem. A* **2016**, *5*, 2411–2428. [CrossRef]
11. Liu, Y.; Lv, B.; Liu, P.; Chen, Y.; Gao, B.; Lin, B. Biotemplate-assisted hydrothermal synthesis of tubular porous Co$_3$O$_4$ with excellent charge-discharge cycle stability for supercapacitive electrodes. *Mater. Lett.* **2018**, *210*, 231–234. [CrossRef]
12. Cakici, M.; Kakarla, R.R.; Alonso-Marroquin, F. Advanced electrochemical energy storage supercapacitors based on the flexible carbon fiber fabric-coated with uniform coral-like MnO$_2$ structured electrodes. *Chem. Eng. J.* **2017**, *309*, 151–158. [CrossRef]
13. Sun, L.; Tian, C.G.; Li, M.T.; Meng, X.Y.; Wang, L.; Wang, R.H.; Yin, J.; Fu, H.G. From coconut shell to porous graphene-like nanosheets for high-power supercapacitors. *J. Mater. Chem. A* **2013**, *1*, 6462–6470. [CrossRef]
14. Saravanan, K.R.; Kalaiselvi, N. Nitrogen containing bio-carbon as a potential anode for lithium batteries. *Carbon* **2015**, *81*, 43–53. [CrossRef]
15. Tong, G.; Liu, Y.; Guan, J. In situ gas bubble-assisted one-step synthesis of polymorphic Co$_3$O$_4$ nanostructures with improved electrochemical performance for lithium ion batteries. *J. Alloys Compd.* **2014**, *601*, 167–174. [CrossRef]
16. Sun, D.Y.; He, L.W.; Chen, R.Q.; Lin, Z.Y.; Lin, S.S.; Xiao, C.X.; Lin, B.Z. The synthesis, characterization and electrochemical performance of hollow sandwich microtubules composed of ultrathin Co$_3$O$_4$ nanosheets and porous carbon by bio-template. *J. Mater. Chem. A* **2018**, *6*, 18987–18993. [CrossRef]
17. Sun, D.Y.; He, L.W.; Chen, R.Q.; Liu, Y.; Lv, B.J.; Lin, S.S.; Lin, B.Z. Biomorphic composites composed of octahedral Co$_3$O$_4$ nanocrystals and mesoporous carbon microtubes templated from cotton for excellent supercapacitor electrodes. *Appl. Surf. Sci.* **2019**, *465*, 232–240. [CrossRef]
18. Wang, L.; Duan, G.R.; Chen, S.M.; Liu, X.H. Hydrothermally controlled synthesis of α-MnO$_2$, γ-MnOOH, and Mn$_3$O$_4$ nanomaterials with enhanced electrochemical properties. *J. Alloys Compd.* **2018**, *752*, 123–132. [CrossRef]
19. Zhou, J.; Zhao, H.; Mu, X.; Chen, J.; Zhang, P.; Wang, Y.; He, Y.; Zhang, Z.; Pan, X.; Xie, E. Importance of polypyrrole in constructing 3D hierarchical carbon nanotube@MnO$_2$ perfect core-shell nanostructures for high-performance flexible supercapacitors. *Nanoscale* **2015**, *7*, 14697–14706. [CrossRef] [PubMed]
20. Zhu, J.; Xu, Y.; Hu, J.; Wei, L.; Liu, J.; Zheng, M. Facile synthesis of MnO$_2$ grown on nitrogen-doped carbon nanotubes for asymmetric supercapacitors with enhanced electrochemical performance. *J. Power Sources* **2018**, *393*, 135–144. [CrossRef]

© 2018 by the authors. Licensee MDPI, Basel, Switzerland. This article is an open access article distributed under the terms and conditions of the Creative Commons Attribution (CC BY) license (http://creativecommons.org/licenses/by/4.0/).

Article

Super-Capacitive Performance of Manganese Dioxide/Graphene Nano-Walls Electrodes Deposited on Stainless Steel Current Collectors

Roger Amade [1,2,*], Arevik Muyshegyan-Avetisyan [1,2], Joan Martí González [1,2], Angel Pérez del Pino [3], Eniko György [3], Esther Pascual [1,2], José Luís Andújar [1,2] and Enric Bertran Serra [1,2]

1. ENPHOCAMAT (FEMAN) Group, Department of Applied Physics, Universitat de Barcelona, Martí i Franquès 1, E-08028 Barcelona, Spain; amusheghyan91@ub.edu (A.M.-A.); joanmarti13@gmail.com (J.M.G.); epascual@ub.edu (E.P.); jandujar@ub.edu (J.L.A.); ebertran@ub.edu (E.B.S.)
2. Institute of Nanoscience and Nanotechnology (IN2UB), Universitat de Barcelona, E-08028 Barcelona, Spain
3. Instituto de Ciencia de Materiales de Barcelona, Consejo Superior de Investigaciones Científicas (ICMAB-CSIC), Campus UAB, E-08193 Bellaterra, Spain; aperez@icmab.es (A.P.d.P.); egyorgy@icmab.es (E.G.)
* Correspondence: r.amade@ub.edu; Tel.: +34-93-403-7089

Received: 28 December 2018; Accepted: 30 January 2019; Published: 4 February 2019

Abstract: Graphene nano-walls (GNWs) are promising materials that can be used as an electrode in electrochemical devices. We have grown GNWs by inductively-coupled plasma-enhanced chemical vapor deposition on stainless steel (AISI304) substrate. In order to enhance the super-capacitive properties of the electrodes, we have deposited a thin layer of MnO_2 by electrodeposition method. We studied the effect of annealing temperature on the electrochemical properties of the samples between 70 °C and 600 °C. Best performance for supercapacitor applications was obtained after annealing at 70 °C with a specific capacitance of 104 $F \cdot g^{-1}$ at 150 $mV \cdot s^{-1}$ and a cycling stability of more than 14k cycles with excellent coulombic efficiency and 73% capacitance retention. Electrochemical impedance spectroscopy, cyclic voltammetry, and galvanostatic charge/discharge measurements reveal fast proton diffusion (1.3×10^{-13} $cm^2 \cdot s^{-1}$) and surface redox reaction after annealing at 70 °C.

Keywords: inductively-coupled plasma; carbon nanostructures; electrochemical properties; thermal annealing

1. Introduction

Carbon-based nanostructured electrodes are suitable for different electrochemical applications due to their large surface area, thermal, and chemical stability, mechanical strength, and high conductivity. Carbon nanotubes (CNTs), graphene, and graphene nano-walls (GNWs) are promising materials to improve performance of different electrochemical devices such as sensors, batteries, electrochemical capacitors, and fuel cells [1–6].

Among the different carbon nanostructures, graphene nano-walls is one of the most recently developed and less investigated material. GNWs can be grown by different plasma enhanced chemical vapor deposition (PECVD) techniques such as microwave PECVD (MW-PECVD) [7], capacitively-coupled radio-frequency PECVD (CC-rf-PECVD) [8], direct current PECVD (DC-PECVD) [9], and inductively coupled plasma enhanced chemical vapor deposition (ICP-CVD) [10]. ICP-CVD is one of the high plasma density and large plasma volume techniques, which allow synthesizing GNWs at high deposition rates. In addition, ICP-CVD has a simple geometry and operates at relatively low pressures. There are two geometric designs for ICP reactor known as planar and cylindrical coil geometry [11]. The main method used in this work is the ICP-CVD method with a cylindrical coil geometry and substrate at floating potential inside the plasma.

ICP-CVD allows the growth of GNWs on different substrates without the need of a diffusion barrier nor a catalyst typically used for the growth of CNTs. Consequently, the synthesis process is greatly simplified. The growth mechanism of GNWs is still under discussion, but the first stage seems to be the deposition of an amorphous carbon layer on the substrate, which acts as a catalyst for the growth of the nano-walls [1].

Carbon-based nanostructured electrodes such as GNWs are especially promising for supercapacitor devices. These electrochemical devices present higher power density values than batteries as well as higher energy density values than electrolytic capacitors. Supercapacitors are used and are under development to improve the performance of electric vehicles, memory back-ups, and absorption of high peak power transients. Recently, several articles have shown the promising properties of GNWs as electrodes for electric double layer capacitors (EDLC). Researchers have been able to grow them on metallic substrates [12–14], flexible substrates [15], controlling their surface functionalization [16], their morphology [17], and obtaining high voltage energy storage devices [18]. However, very few articles have explored the electrochemical properties of metal oxide/GNWs nanocomposites for energy storage devices [19–21].

There are two main charging mechanisms in a super-capacitor such as electrostatic due to the formation of an electric double layer at the interface between the solid electrode and the ions in the electrolyte (double-layer capacitance), and through redox reactions near the surface region of transition metal oxides or conductive polymers involving electrolyte ions (pseudo-capacitance). In general, carbon electrodes have a higher contribution of double layer capacitance (DLC) than pseudo-capacitance (PC) [22]. Due to their rich redox behavior, thermal and chemical stability, transition metal oxides provide higher energy density than conventional carbon materials and better electrochemical stability than polymer materials [23].

Among super-capacitive transition metal oxides, manganese dioxide is one of the most investigated due to its environmental friendliness, cost, and low toxicity [23]. The proposed mechanism for pseudo-capacitance in manganese oxides is described by the following equation [24].

$$MnO_\alpha(OC)_\beta + \delta C^+ + \delta e^- \leftrightarrow MnO_{\alpha-\delta}(OC)_{\beta+\delta} \qquad (1)$$

where C^+ are protons or alkali metal cations (Li$^+$, Na$^+$, K$^+$) in the electrolyte. In the case of adsorption/desorption of protons, $MnO_\alpha(OH)_\beta$ and $MnO_{\alpha-\delta}(OH)_{\beta+\delta}$ indicate interfacial manganese hydroxides at high and low oxidation states, respectively. A major drawback of manganese dioxide is its low conductivity that reduces the charging/discharging rate of the electrode. Thus, a thin layer of the oxide is deposited to minimize the series resistance. In addition, a thermal treatment may help improve the contact between the transition metal oxide and the carbon-based nanostructured electrode improving the charge transfer process [25]. Optimization of this thermal treatment allows maximum charge/discharge rates without sacrificing electrochemical performance.

Different methods can be used to deposit and prepare MnO_2, such as hydrothermal [26], sol gel methods [24,27], and electrodeposition [24,27,28]. In previous studies, we have demonstrated the excellent electrochemical performance of galvanostatically electrodeposited MnO_2 on vertically aligned carbon nanotubes (VACNTs) [28,29] as well as laser-coated MnO_2-VACNTs [30]. In addition, there is a need to reduce overall costs of electrochemical devices by the use of cost-effective current collectors such as stainless-steel foil. This substrate has the advantages of being flexible, relatively chemically stable, and with low resistance.

In this case, we have grown GNWs on stainless steel AISI304 (SS) by the ICP-CVD method, which electrodeposited a thin layer of MnO_2. Afterward, we have performed a thermal annealing process to improve the electrochemical properties of the electrodes and analyzed their performance. A mild thermal treatment at 70 °C resulted in ultrafast charge/discharge rates without giving up performance and stability.

2. Materials and Methods

2.1. Growth of GNWs and MnO$_2$/GNWs

Graphene nano-walls were synthesized on the SS substrate (100 µm thick and about 2 cm^2 geometrical area) using an inductively coupled plasma chemical vapor deposition (ICP-CVD) (13.56 MHz, power 440 W) system [31] consisting on a long quartz tube (Vidrasa S.A., Ripollet, Spain) having a radio-frequency resonator (homemade), for producing remote plasma, and a tubular oven (PID Eng & Tech S.L., Madrid, Spain) working up to 1100 °C, 20 cm away from the quartz tube (Figure 1).

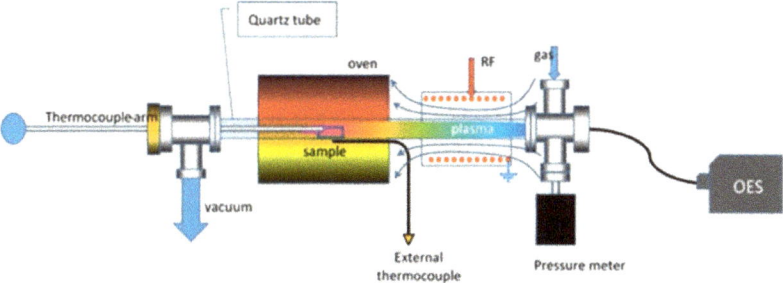

Figure 1. Schematic drawing of the ICP-CVD tubular reactor for producing carbon structures. The reactor is composed of a quartz tube, an oven, and a coil connected to an RF power supply. The sample is placed in the middle of the oven. The system is evacuated with a primary rotatory pump and the pressure were analyzed with a pressure meter. The temperature is measured by means of two thermocouples. Different gases can be introduced through an inlet at one end of the quartz tube. OES: Optical Emission Spectroscopy device to study the emission of plasma generated species.

The sample was placed inside the quartz tubular reactor, which was evacuated down to a pressure below 1 Pa and heated up to 750 °C. Then, pure methane (99.995%) was introduced as a precursor gas at one end of the quartz tube (10 sccm) and the pressure was maintained in the range of 50 to 60 Pa. Under these conditions, plasma was ignited at an RF power of 440 W for 40 min. Lastly, the sample was cooled down to room temperature for 30 min. The areal density of the GNWs was measured with a microbalance and found to be about 58 µg·cm^{-2}.

The electrochemical deposition was carried out using a galvanostatic method described elsewhere [28]. GNWs were used as anode and a graphite electrode as the cathode in a 0.2 M Na$_2$SO$_4$ aqueous solution with about 4 cm separation between electrodes. About 0.5 cm^3 of a 0.2 M MnSO$_4$·H$_2$O solution was added drop wise to the electrolyte through a hole in the cathode applying 1 mA·cm^{-2} for 2 min [32]. After this process and previous to the heat treatment, the loading mass of MnO$_2$ was determined by UV–vis spectrophotometry (PerkinElmer Inc., Shelton, CT, USA). First, MnO$_2$ was dissolved in concentrated nitric acid. Then, manganese ions were oxidized to MnO$_4^-$ using K$_2$S$_2$O$_8$ and AgNO$_3$. From the characteristic absorbance maximum at 525 nm, the concentration of MnO$_4^-$ could be determined. From this value, the mass loading of MnO$_2$ was calculated, which was about 16 µg [32] (see supplementary information, Figure S1).

The MnO$_2$/GNWs samples were annealed at different temperatures to study their electrochemical performance: 70 °C, 200 °C, and 600 °C. The annealing process was carried out for 20 min under an argon atmosphere to avoid oxidation of the SS substrate. MnO$_2$/GNWs without annealing (room temperature) and bare GNWs samples were also characterized for a comparison.

2.2. Samples Characterization

The microstructure and morphologies of the grown samples were investigated using field emission scanning electron microscopy (SEM) (JEOL JSM-7001F, operated at 20 kV, JEOL Ltd., Tokyo, Japan) and transmission electron microscopy (TEM) (JEOL 1010, operated at 200 kV, JEOL Ltd.). The carbon structures were transferred to the TEM grid by simply scratching off from the substrates.

XPS (X-ray photoelectron spectroscopy) experiments were performed in a PHI 5500 Multi-Technique System (Physical Electronics, Chanhassen, MN, USA) with a monochromatic X-ray source (Aluminium Kalfa line of 1486.6 eV energy and 350 W), placed perpendicular to the analyzer axis, and calibrated using the 3d5/2 line of Ag with a fullwidth at half maximum (FWHM) of 0.8 eV. The analyzed area was a circle of 0.8 mm diameter, and the selected resolution for the spectra was 187.5 eV of pass energy and 0.8 eV/step for the general spectra and 11.75 eV of pass energy and 0.05 eV/step for the spectra of the different elements in the depth profile spectra. All measurements were made in an ultra-high vacuum (UHV) chamber pressure between 5×10^{-9} and 2×10^{-8} torr.

The physical and chemical characteristics of, as grown structures, were studied by Raman spectroscopy using a microscope HR800 (LabRam) (HORIBA France SAS, Palaiseau, France) with a 532 nm solid-state laser (5 mW laser power).

The electrochemical properties of the samples were analyzed by using cyclic voltammetry (CV), electrochemical impedance spectroscopy (EIS), and galvanostatic charge/discharge (GCD) cycling in a 1 M Na_2SO_4 aqueous solution using a potentiostat/galvanostat (AutoLab, PGSTAT30, Eco Chemie B.V., Utrecht, The Netherlands). All experiments were carried out at room temperature in a typical three-electrode cell. A Ag/AgCl electrode (3 M KCl internal solution) and a Pt-ring electrode were used as the reference and counter electrode, respectively. The working electrode was a sample of GNWs or MnO_2/GNWs nanocomposite. The geometrical area of the working electrode was set to a constant value of 0.57 cm^2.

The CV analysis was performed using a voltage window of 0.0 to 0.7 V vs. Ag/AgCl at scan rates between 10 and 150 $mV \cdot s^{-1}$. An alternating voltage of 10 mV amplitude was applied to the samples between 1 Hz and 100 kHz for the EIS analysis. Lastly, several thousands of charge/discharge cycles were applied to the samples between 0.7 V and 0.0 V vs. Ag/AgCl at different current densities.

3. Results

3.1. Morphological and Structural Characterization

Figure 2a shows a top view image of bare GNWs grown on SS. The nano-walls are homogeneously distributed and present an open structure with voids between the nano-walls. Their length is of several hundreds of nanometers and a few nanometer thick (about 10 graphene layers, see Figure 2e,f).

After deposition of MnO_2, a layer of the oxide is observed coating the nano-walls. The presence of manganese and oxygen was determined by EDX (energy dispersive X-ray spectroscopy) and XPS [29] (see Figures S2 and S3 in supplementary information for more details). For samples without thermal treatment and annealed at 200 °C or below (Figure 2b,c, respectively), the open structure of the nano-walls is still recognized. However, a completely different morphology is presented by the sample annealed at 600 °C (Figure 2d) with a homogeneous layer of the oxide covering the nano-walls without visible voids and with a more compact structure.

Raman spectra of bare GNWs show typical D and G bands at 1346 cm^{-1} and 1587 cm^{-1} (Figure 3). The large number of defects (high D-peak) is associated with the plasma deposition technique. A shoulder appears next to the G band at 1616 cm^{-1} (D' band) related to the finite size of graphite crystals and graphene edges [33]. Furthermore, the presence of a peak at 2700 cm^{-1} (2D) and $I_{2D}/I_G = 0.8$ confirms the presence of few-layered graphene in accordance with TEM images (Figure 2e,f). An additional peak appears at 2940 cm^{-1} related to sp^2 CH stretching vibrations [34]. MnO_2/GNWs samples without and with annealing at different temperatures show similar Raman spectra, which indicate little or no alteration of the graphene nano-walls structure after MnO_2 deposition and thermal treatment (Figure 3). Main D and G bands are clearly recognized for all the samples without big differences between them. However, unlike the bare GNWs sample, the D' shoulder is not appreciable for these samples. In addition, 2D and D + G bands appear much weaker than for bare GNWs sample due to the presence of MnO_2 coating on the surface and edges of the nano-walls. No specific bands of MnO_2 are recorded, likely due to its low loading, low crystallinity, and/or small crystal size [30].

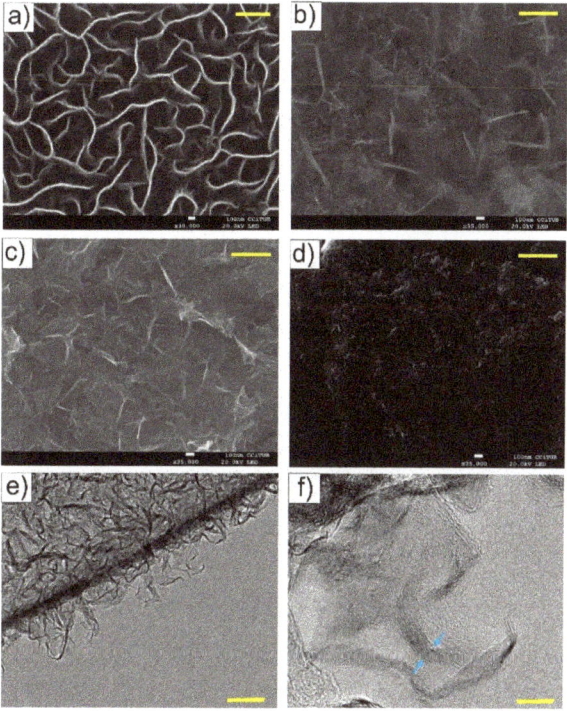

Figure 2. SEM images of GNWs grown on SS, (**a**) MnO_2/GNWs without thermal treatment, (**b**) MnO_2/GNWs with annealing at 200 °C (**c**) and 600 °C (**d**). TEM images of as grown GNWs (**e**), (**f**). Arrows indicate thickness of the nano-walls. Scale bars correspond to 500 nm in (**a–d**). 100 nm in (**e**) and 10 nm in (**f**).

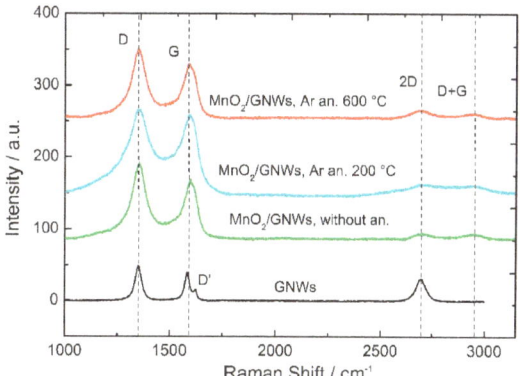

Figure 3. Raman spectra of GNWs grown on SS and MnO_2/GNWs samples without annealing and Argon annealing at 200 and 600 °C.

3.2. Electrochemical Performance

The average specific capacitance of the samples was obtained from the cyclic voltammograms applying Equation (2).

$$C_s = \frac{q_a + |q_c|}{2m\Delta V} \quad (2)$$

where C_s is the average specific capacitance in $F \cdot g^{-1}$, q_a and q_c are the anodic and cathodic charges, respectively in C. m is the mass of MnO_2/GNWs composite in g and ΔV the applied voltage window in V.

Alternatively, the specific capacitance was also obtained from galvanostatic discharge curves, according to the following equation.

$$C_s = \frac{I}{(\Delta V/\Delta t)m} \quad (3)$$

where C_s and m have the same meaning and units, as previously shown (Equation (2)). ΔV is the voltage difference during the discharge in V and Δt the discharge time in s. I is the current applied during discharge in A.

Cyclic voltammograms of the samples exhibit a rectangular shape that indicates the capacitive nature of the GNWs and the MnO_2/GNWs nanocomposites (Figure 4a,b). Deposition of MnO_2 increases the current density of MnO_2/GNWs by a factor greater than 20 compared to bare GNWs samples due to the pseudocapacitive effect of the transition metal oxide (Figure 4b).

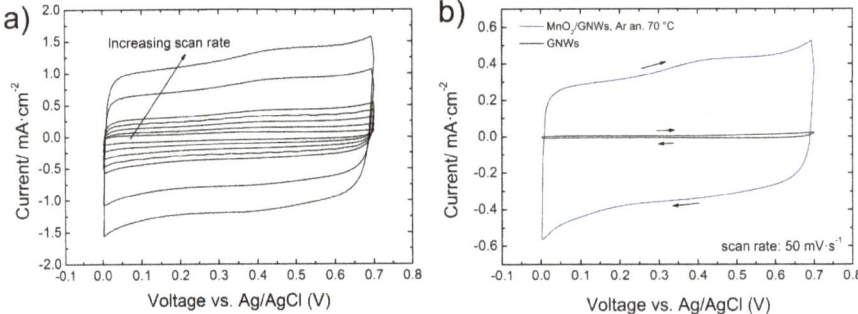

Figure 4. (**a**) Cyclic voltammograms of MnO_2/GNWs sample annealed at 70 °C using scan rates 10, 20, 30, 40, 50, 100, and 150 $mV \cdot s^{-1}$. (**b**) Bare GNWs and MnO_2/GNWs sample annealed at 70 °C using a scan rate of 50 $mV \cdot s^{-1}$ and a voltage window between 0.0 to 0.7 V vs. Ag/AgCl. Arrows indicate the direction of the scan.

At a scan rate of 10 $mV \cdot s^{-1}$, MnO_2/GNWs sample without annealing provides the highest specific capacitance (SC) of 133 $F \cdot g^{-1}$ (Figure 5). However, at high scan rates (150 $mV \cdot s^{-1}$), the highest SC is presented by MnO_2/GNWs sample annealed at 70 °C (104 $F \cdot g^{-1}$). Galvanostatic charge/discharge measurements are in agreement with these results. Figure 6 shows the galvanostatic charge/discharge curves for the sample annealed at 70 °C. The symmetrical triangular shape of the curves indicate excellent coulombic efficiency (>97%). The capacitance increases slightly with the current density (see inset graph in Figure 6), as observed from CV measurements (Figure 5). In addition, this sample also shows the highest cycling stability at high current densities (5 $A \cdot g^{-1}$) after 10,000 cycles and excellent cycling stability at low current densities (1 $A \cdot g^{-1}$) (Figure 7). The sample without thermal treatment shows a capacitance retention of only 50% at 3 $A \cdot g^{-1}$ after 10k cycles, and the rest of the samples present even poorer performance in comparison to the sample annealed at 70 °C.

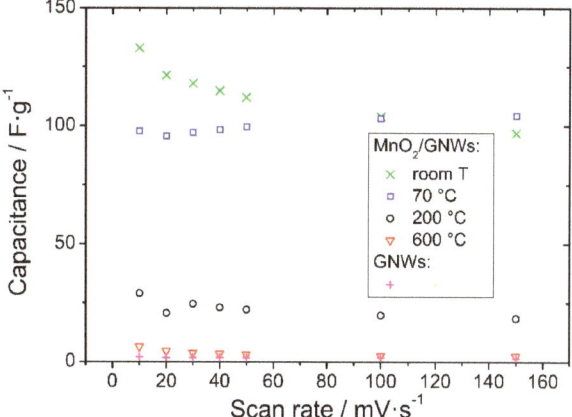

Figure 5. Specific capacitance of bare GNWs and MnO$_2$/GNWs samples without (room T) and with annealing at different temperatures versus the scan rate.

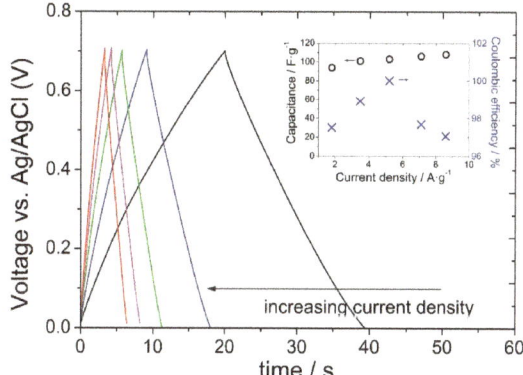

Figure 6. Galvanostatic charge/discharge curves of MnO$_2$/GNWs sample annealed at 70 °C. The values obtained agree with those from CV measurements and show excellent coulombic efficiency (>97%) at high current densities (see inset graph), which were indicated by the symmetry of the charge/discharge curves.

3.3. Impedance Spectroscopy

Further information about the electrochemical processes taking place at the interfaces and bulk of the electrode materials can be obtained by means of electrochemical impedance spectroscopy.

The Nyquist plot of the samples is in accordance with their morphology and show the typical behavior for capacitive porous electrodes (Figure 8). A modified Randles circuit can be used to fit the experimental data of MnO$_2$/GNWs [35] (Figure 8), which describe charge storage and transfer properties of the different interfaces and bulk of the electrode/electrolyte system. The equivalent circuit consist of capacitances (C) and resistances (R) in series and/or parallel configuration. The internal resistance of the cell is given by R_S, which corresponds to the intersection point with the real axis at high frequencies in the Nyquist spectra (inset in Figure 8). The charge transfer resistance between the electrode and the electrolyte is represented by R_{CT}. The double layer capacitance connected in parallel is given by C_{DL}. In the mid-frequency region of the spectrum, a Warburg element (W_O) describes diffusion processes of the ions through the porous structure of the electrodes, which is expressed as $A/(j\omega)^n$, where A is the Warburg coefficient, n is an exponent, and ω is the angular frequency. At low frequencies, the spectra show an almost straight line parallel to the imaginary axis, which is related to a perfect polarized capacitive behavior described by mass capacitance (C_L), which is connected in parallel to a leakage resistance (R_L) [36]. For the GNWs sample, a more simple circuit was used, which was composed of a resistance (R_S) in series with a constant phase element describing an almost ideal capacitive behavior.

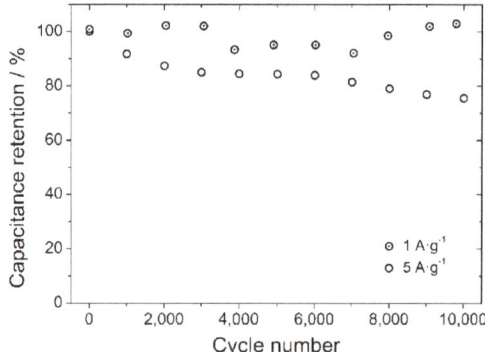

Figure 7. Cycling stability at different current densities (1 and 5 $A \cdot g^{-1}$) of MnO$_2$/GNWs sample annealed at 70 °C. At 1 $A \cdot g^{-1}$, the capacitance retention remains almost constant and around 100% throughout 10k cycles. At 5 $A \cdot g^{-1}$, capacitance retention decreases down to 75% after 10k cycles and the coulombic efficiency is >98% throughout the whole measurement for both current densities.

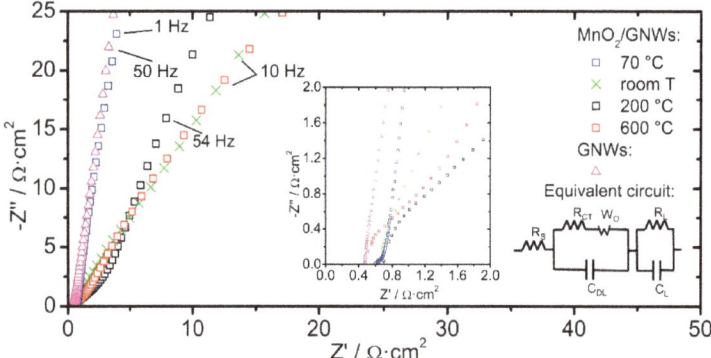

Figure 8. Nyquist plot of the samples: bare GNWs and MnO$_2$/GNWs nanocomposite without annealing (room T) and annealed at 70 °C, 200 °C, and 600 °C. Inset shows real axis intercection and the modified Randles equivalent circuit used to fit the data of MnO$_2$/GNWs.

The whole spectra (1 Hz to 100 kHz) were fitted using ZVIEW software (Version 2.1, Scribner Associates, Inc., Southern Pines, NC, USA) and the equivalent circuit parameters obtained for MnO$_2$/GNWs are shown in Table 1. GNWs presented a series resistance of 0.4 $\Omega \cdot cm^2$ and a double layer capacitance of about 5 $F \cdot g^{-1}$, in accordance with CV measurements.

Table 1. Equivalent circuit parameters obtained from fitting the EIS data of MnO$_2$/GNWs nanocomposites.

Electrode Material/ Annealing T (°C)	R_S ($\Omega \cdot cm^2$)	R_{CT} ($\Omega \cdot cm^2$)	$W_0 = A/(j\omega)^n$ A ($\Omega \cdot s^{-n}$)	n	C_{DL} (F·g^{-1})	R_L ($\Omega \cdot cm^2$)	C_L (F·g^{-1})
MnO$_2$/GNWs/-	0.7	0.9	490	0.54	6.7	637	206
MnO$_2$/GNWs/70	0.6	0.1	8.52	0.65	25	195	231
MnO$_2$/GNWs/200	0.7	1.5	101	0.60	0.63	124	14
MnO$_2$/GNWs/600	0.5	0.6	923	0.61	0.84	4.3	187

Although ions from the bulk solution are transported throughout the porous electrode, it is known that proton diffusion inside the MnO$_2$ lattice is a key step during charge/discharge [37]. Since the concentration of Na$_2$SO$_4$ is high, it is assumed that the diffusion of electrolyte ions through the porous electrode is fast in comparison to the diffusion of protons through the MnO$_2$ lattice and, therefore, proton diffusion is the rate-determining step in the low frequency region. Thus, the proton diffusion coefficient (D) can be obtained from the slope of the Z' vs. $\omega^{1/2}$ plot (Figure 9b) through the following equation [38,39].

$$D = \frac{R^2 T^2 M^2}{2 A^2 n^4 F^4 \sigma^2} \quad (4)$$

where R is the gas constant in J·mol^{-1}·K^{-1}, T is the temperature in K, M is the molar volume (taken as 17.30 cm^3·mol^{-1} for MnO$_2$), A is the geometric area of the electrode in cm^2, n is the number of electrons involved in the reaction, F is the Faraday constant in C·mol^{-1}, and σ is the slope of Figure 9b in $\Omega \cdot s^{-1/2}$. The maximum proton diffusion coefficient is obtained for the sample annealed at 70 °C with a value of 1.3×10^{-13} cm$^2 \cdot s^{-1}$.

The relaxation time constant (τ_0) can be obtained from the C" vs. frequency plot (Figure 9a) using Equation (5) [36,40].

$$C'' = \frac{Z'}{2\pi f |Z|^2} \quad (5)$$

where C" is the imaginary part of the capacitance in F·cm^2, Z' is the real part of the impedance in $\Omega \cdot cm^2$, f is the frequency in Hz, and |Z| is the modulus of the impedance in $\Omega \cdot cm^2$. The position of the maximum in Figure 9a provides the characteristic frequency, which is the inverse of τ_0. The fastest relaxation time was 25 ms, which is obtained for the sample annealed at 70 °C.

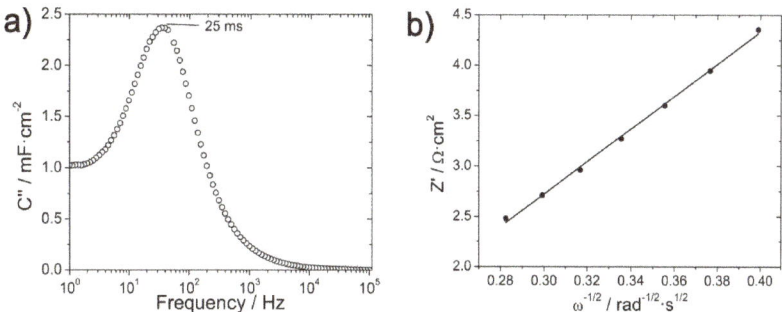

Figure 9. (a) C" vs. frequency plot to determine the relaxation time constant; (b) Z' vs. $\omega^{1/2}$ plot to evaluate the diffusion coefficient of MnO$_2$/GNWs heated at 70 °C.

4. Discussion

The heat-treated samples are expected to show lower open porosity and lower surface area, especially those heated at high temperatures above 200 °C. In addition, it has been suggested that high scan rates limits proton and Na$^+$ ion diffusion and makes some pores and voids inaccessible [41], which implies a decrease in SC with a scan rate. It is also known that a heat treatment below 220 °C removes water and trace amounts of oxygen [42], and, around 500 °C, a transition occurs from α-MnO$_2$ to crystalline Mn$_2$O$_3$ by loss of oxygen (chemical transformation) [42–44], which explains the low SC of the samples treated above 200 °C. However, the sample annealed at 70 °C shows a particular behavior increasing its SC with the scan rate (Figure 5 and inset graph in Figure 6). This phenomenon has already been reported previously and has been assigned to an excellent contact between the porous collector (GNWs in this case) and the oxide that allows efficient access of both electrons and ions to afford a fast redox reaction at high scan rates [45]. This redox process takes place at the surface of MnO$_2$ where only a thin layer of MnO$_2$ is involved and is electrochemically active [46].

Clearly, high annealing temperatures drastically reduce the capacitance due to the removal of bound water and lower open porosity of the electrode [41]. A mild treatment at 70 °C improves the SC and provides a better cycling stability even at high current densities (Figure 7). The Coulombic efficiency for this sample remained above 98% even after 10 k cycles at 1 and 5 A g^{-1}, and the capacitance retention was excellent at 1 A·g^{-1} and 75% at 5 A·g^{-1}.

The cycling stability of manganese oxides is mainly controlled by their microstructure (physical property), while their specific capacitance is governed primarily by their chemically hydrous state [47]. Thus, when the annealing temperature is too high (above 200 °C), the hydrous state is reduced and, without annealing or at low temperatures, the microstructure is not as stable as after a heat treatment. Optimum conditions are found at 70 °C annealing temperature.

As expected, the series resistance values are similar to all the samples, which are related to the electrolyte resistance, cables, and contact resistances of the cell [12] (Table 1). Due to the low conductivity of MnO$_2$, the R$_S$ value increases slightly after its deposition. However, after heating the sample at 600 °C, this value decreases again, probably due to detachment of the oxide layer from the GNWs exposing the nanocarbon material to the electrolyte. The charge transfer resistance presents its lowest value after annealing the MnO$_2$/GNWs at 70 °C. It is assumed that this mild temperature allows better contact of the oxide with the GNWs and avoids fast water evaporation that can result in big voids, cracks, and bad adhesion between oxide and GNWs. This is probably the case for the sample annealed at 200 °C. At higher temperatures, the R$_{CT}$ decreases again due to the increased diffusivity of the metal-oxide ions during annealing that result in better contact between GNWs and MnO$_2$. Both double layer capacitance and mass capacitance have its maximum after annealing at 70 °C, which correlates with the cyclic voltammetry results. The leakage resistance diminishes with an increasing annealing temperature. The origin of this resistance is usually related to some faradaic process [22], and may be related to chemical reactions or phase changes of the manganese oxide. At annealing temperatures below 500 °C, there is a loss of water and a slight loss of oxygen [48], while a phase change from α-MnO$_2$ to Mn$_2$O$_3$ takes place above 500 °C [47]. Lower resistance implies higher leaking currents and poor performance. Hence, low annealing temperatures should be used. The Warburg coefficient presents its lowest value for the sample annealed at 70 °C. The exponent n is usually 0.5 for porous carbon electrodes. The high value of 0.65 obtained for the 70 °C annealed sample signifies that ion diffusion takes place only at the surface of the electrode and does not behave as a typical porous electrode [35]. This result is in accordance with the proton diffusion coefficient (D) obtained from the slope of the Z' vs. $\omega^{1/2}$ plot (Figure 9a).

In agreement with previously reported values [37], the obtained proton diffusion coefficients in manganese dioxide range from 3.3×10^{-16} to 2.4×10^{-15} cm^2·s^{-1} for all the samples except for the one annealed at 70 °C, which presents a value of 1.3×10^{-13} cm^2·s^{-1}. The fast diffusion of protons in this sample is in accordance with the low Warburg coefficient and high n value obtained (Table 1).

The relaxation time is related to the power delivery and, ideally, should be as small as possible. The shortest relaxation time is obtained after annealing at 70 °C (Figure 9a), which facilitates fast rate performance during charge/discharge cycles.

Hassan et al. [21] also measured the capacitive properties of MnO$_2$/carbon nano-walls (CNWs) composite on nickel foam, and obtained an SC of about 310 F·g^{-1} at 150 mV·s^{-1} considering the mass of MnO_2 only. In this work, we obtained an SC of 278 $F \cdot g^{-1}$ (considering the mass of MnO_2 only) at 150 $mV \cdot s^{-1}$ for the sample annealed at 70 °C. The two values are similar, but, in this case, we used SS as a substrate, which is more cost-effective.

In summary, graphene nano-walls have been successfully grown on conductive and flexible stainless steel substrate by ICP-CVD method. Galvanostatic electrodeposition of MnO_2 increases the SC of the samples up to 133 $F \cdot g^{-1}$ at 10 $mV \cdot s^{-1}$ for the sample without thermal treatment. However, after a mild annealing at 70 °C, the MnO_2/GNWs nanocomposite exhibits optimum electrochemical performance, i.e., small charge transfer resistance, high cycling stability, excellent coulombic efficiency, and SC even at high charge/discharge rates, which is related to optimum contact between GNWs and MnO_2 and fast surface redox reactions. In accordance with these results, a maximum in the proton diffusion coefficient and short relaxation time is found at 70 °C annealing temperature, which suggests fast surface diffusion processes. Samples annealed above 200 °C show low SC and stability due to the reduced open porosity, bound water removal from manganese oxide, and faradaic reactions or phase changes.

Supplementary Materials: The following are available online at http://www.mdpi.com/1996-1944/12/3/483/s1, Figure S1: Calibration curve for the determination of manganese dioxide in the samples. The absorbance maximum at 525 nm is plotted against the concentration of MnO_4^- in aqueous solution, Figure S2: Energy dispersive x-ray spectrum of a MnO2/GNWs composite sample after galvanostatic electrodeposition of MnO_2, Figure S3: XPS spectra of galvanostatically electrodeposited MnO_2 without thermal treatment.

Author Contributions: Conceptualization, R.A., A.M.-A., J.M.G., A.P.d.P., E.G., E.P., J.L.A. and E.B.S.; Methodology, R.A., A.M.-A., J.M.G. and E.B.S.; Software, R.A., A.M.-A., J.M.G., A.P.d.P., E.G. and E.B.S.; Validation, R.A., A.M.-A. and J.M.G.; Formal Analysis, R.A., A.M.-A., J.M.G., A.P.d.P and E.G.; Investigation, R.A., A.M.-A., J.M.G., A.P.d.P., E.G. and E.B.S.; Resources, A.P.d.P., E.G., E.P., J.L.A. and E.B.S.; Data Curation, R.A., A.M.-A. and J.M.G.; Writing—Original Draft Preparation, R.A. and A.M.-A.; Writing—Review And Editing, R.A., A.M.-A., J.M.G., A.P.d.P., E.G., E.P., J.L.A. and E.B.S.; Visualization, R.A., A.M.-A. and J.M.G.; Supervision, R.A., A.P.d.P., E.G., E.P., J.L.A. and E.B.S.; Project Administration, R.A., A.P.d.P., E.G., E.P., J.L.A. and E.B.S.; Funding Acquisition, A.P.d.P., E.G., E.P., J.L.A. and E.B.S.

Funding: This research was funded by the Spanish Ministry of Economy, Industry and Competitiveness, grant numbers ENE2014-56109-C3-1-R, ENE2014-56109-C3-3-R, ENE2017-89210-C2-1-R, and ENE2017-89210-C2-2-R, and from AGAUR of Generalitat de Catalunya, grant number 2017SGR1086. The article processing charges were funded by the Spanish Ministry of Economy, Industry and Competitiveness, grant number EUIN2017-88755.

Acknowledgments: Two authors (A.M-A and J.M-G) acknowledge the financial support from their respective APIF grants from the Universitat de Barcelona.

Conflicts of Interest: The authors declare no conflict of interest.

References

1. Bo, Z.; Mao, S.; Han, Z.J.; Cen, K.; Chen, J.; Ostrikov, K. Emerging energy and environmental applications of vertically-oriented graphenes. *Chem. Soc. Rev.* **2015**, *44*, 2108–2121. [CrossRef] [PubMed]
2. Wen, L.; Li, F.; Cheng, H.-M. Carbon nanotubes and graphene for flexible electrochemical energy storage: From materials to devices. *Adv. Mater.* **2016**, *28*, 4306–4337. [CrossRef] [PubMed]
3. Baptista, F.R.; Belhout, S.A.; Giordani, S.; Quinn, S.J. Recent developments in carbon nanomaterial sensors. *Chem. Soc. Rev.* **2015**, *44*, 4433–4453. [CrossRef] [PubMed]
4. Dai, L.; Xue, Y.L.; Qu, L.; Choi, H.-J.; Baek, J.-B. Metal-free catalysts for oxygen reduction reaction. *Chem. Rev.* **2015**, *115*, 4823–4892. [CrossRef] [PubMed]
5. Yang, J.; Wei, D.; Tang, L.; Song, X.; Luo, W.; Chu, J.; Gao, T.; Shi, H.; Du, C. Wearable temperature sensor based on graphene nanowalls. *RSC Adv.* **2015**, *5*, 25609–25615. [CrossRef]
6. Li, B.; Li, S.; Liu, J.; Wang, B.; Yang, S. Vertically aligned sulfur-graphene nanowalls on substrates for ultrafast lithium-sulfur batteries. *Nano Lett.* **2015**, *15*, 3073–3079. [CrossRef] [PubMed]
7. Wu, Y.; Qiao, P.; Chong, T.; Shen, Z. Carbon Nanowalls Grown by Microwave Plasma Enhanced Chemical Vapor Deposition. *Adv. Mater.* **2002**, *14*, 64–67. [CrossRef]
8. Shiji, K.; Hiramatsu, M.; Enomoto, A.; Nakamura, M.; Amano, H.; Hori, M. Vertical growth of carbon nanowalls using rf plasma-enhanced chemical vapour deposition. *Diam. Relat. Mater.* **2005**, *14*, 831–834. [CrossRef]

9. Dikonimos, T.; Giorgi, L.; Lisi, N.; Salernitano, E.; Rossi, R.R. DC plasma enhanced growth of oriented carbon nanowall films by HFCVD. *Diam. Relat. Mater.* **2007**, *16*, 1240–1243. [CrossRef]
10. Hiramatsu, M.; Nihashi, Y.; Kondo, H.; Hori, M. Nucleation control of carbon nanowalls using inductively coupled plasma-enhanced chemical vapor deposition. *Jpn. J. Appl. Phys.* **2013**, *52*, 01AK05. [CrossRef]
11. Sato, G.; Morio, T.; Kato, T.; Hatakeyama, R. Fast growth of carbon nanowalls from pure methane using helicon plasma-enhanced chemical vapour deposition. *Jpn. J. Appl. Phys.* **2006**, *45*, 5210–5212. [CrossRef]
12. Bo, Z.; Wen, Z.; Kim, H.; Lu, G.; Yu, K.; Chen, J. One step fabrication and capacitive behavior of electrochemical double layer capacitor electrodes using vertically-oriented graphene directly grown on metal. *Carbon* **2012**, *50*, 4379–4387. [CrossRef]
13. Zhou, H.; Liu, D.; Luo, F.; Luo, B.; Tian, Y.; Chen, D.; Shen, C. Preparation of graphene nanowalls on nickel foam as supercapacitor electrodes. *Micro Nano Lett.* **2018**, *13*, 842–844. [CrossRef]
14. Nersisyan, H.H.; Lee, S.H.; Choi, J.H.; Yoo, B.U.; Suh, H.; Kim, J.-G.; Lee, J.-H. Hierarchically porous carbon nanosheets derived from alkali metal carbonates and their capacitance in alkaline electrolytes. *Mater. Chem. Phys.* **2018**, *207*, 513–521. [CrossRef]
15. Sahoo, G.; Ghosh, S.; Polaki, S.R.; Mathews, T.; Kamruddin, M. Scalable transfer of vertical graphene nanosheets for flexible supercapacitor applications. *Nanotechnology* **2017**, *28*, 415702. [CrossRef] [PubMed]
16. Deheryan, S.; Cott, D.J.; Mertens, P.W.; Heyns, M.; Vereecken, P.M. Direct correlation between the measured electrochemical capacitance, wettability and surface functional groups of Carbon Nanosheets. *Electrochim. Acta* **2014**, *132*, 574–582. [CrossRef]
17. Ghosh, S.; Polaki, S.R.; Kamruddin, M.; Jeong, S.M.; Ostrikov, K. Plasma-electric field controlled growth of oriented graphene for energy storage applications. *J. Phys. D Appl. Phys.* **2018**, *51*, 145303. [CrossRef]
18. Chi, Y.-W.; Hu, C.-C.; Shen, H.-H.; Huang, K.-P. New approach for high-voltage electric double-layer capacitors using vertical graphene nanowalls with and without nitrogen doping. *Nano Lett.* **2016**, *16*, 5719–5727. [CrossRef]
19. Bai, J.; Yang, L.; Dai, B.; Ding, Y.; Wang, Q.; Han, J.; Zhu, J. Synthesis of $CuO\text{-}Cu_2O$@graphene nanosheet arrays with accurate hybrid nanostructures and tunable electrochemical properties. *Appl. Surf. Sci.* **2018**, *452*, 259–267. [CrossRef]
20. Dinh, T.M.; Acour, A.; Vizireanu, S.; Dinescu, G.; Nistor, L.; Armstrong, K.; Guay, D.; Pech, D. Hydrous RuO_2/carbon nanowalls hierarchical structures for all-solid-state ultrahigh-energy-density micro-supercapacitors. *Nano Energy* **2014**, *10*, 288–294. [CrossRef]
21. Hassan, S.; Suzuki, M.; Mori, S.; Abd El-Moneim, A. MnO_2/carbon nanowalls composite for supercapacitor application. *J. Power Sources* **2014**, *249*, 21–27. [CrossRef]
22. Conway, B.E. *Electrochemical Supercapacitors*; Kluwer Academic Publishers/Plenum Press: New York, NY, USA, 1999; pp. 222, 518, 528.
23. Wang, G.; Zhang, L.; Zhang, J. A review of electrode materials for electrochemical supercapacitors. *Chem. Soc. Rev.* **2012**, *41*, 797–828. [CrossRef] [PubMed]
24. Raymundo-Piñero, E.; Khomenko, V.; Frackowiak, E.; Béguin, F. Performance of manganese oxide/CNTs composites as electrode materials for electrochemical capacitors. *J. Electrochem. Soc.* **2005**, *152*, A229–A235. [CrossRef]
25. Zheng, J.P.; Cygan, P.J.; Jow, T.R. Hydrous ruthenium oxide as an electrode material for electrochemical capacitors. *J. Electrochem. Soc.* **1995**, *142*, 2699–2703. [CrossRef]
26. Tang, N.; Tian, X.K.; Yang, C.; Pi, Z.B. Facile synthesis of a-MnO_2 nanostructures for supercapacitors. *Mater. Res. Bull.* **2009**, *44*, 2062–2067. [CrossRef]
27. Pang, S.C.; Anderson, M.A.; Chapman, T. Novel electrode materials for thin-film ultracapacitors: Comparison of electrochemical properties of sol-gel-derived and electrodeposited manganese dioxide. *J. Electrochem. Soc.* **2000**, *147*, 444–450. [CrossRef]
28. Amade, R.; Jover, E.; Caglar, B.; Mutlu, T.; Bertran, E. Optimization of MnO_2/vertically aligned carbon nanotube composite for supercapacitor application. *J. Power Sources* **2011**, *196*, 5779–5783. [CrossRef]
29. Hussain, S.; Amade, R.; Jover, E.; Bertran, E. Water plasma functionalized CNTs/MnO_2 composites for supercapacitors. *Sci. World J.* **2013**, *2013*, 832581. [CrossRef]
30. Pérez del Pino, A.; György, E.; Alshaikh, I.; Pantoja-Suárez, F.; Andújar, J.-L.; Pascual, E.; Amade, R.; Bertran-Serra, E. Laser-driven coating of vertically aligned carbon nantoubes with manganese oxide from metal organic precursors for energy storage. *Nanotechnology* **2017**, *28*, 395405. [CrossRef]

31. Musheghyan-Avetisyan, A.; Martí González, J.; Bertran Serra, E. Direct growth of vertically oriented graphene nanowalls on multiple substrates by Low Temperature Plasma Enhanced Chemical Vapor Deposition. In Proceedings of the European Graphene Forum 2017 (EGF2017), Paris, France, 26–28 April 2017.
32. Fan, Z.; Chen, J.; Zhang, B.; Sun, F.; Liu, B.; Kuang, Y. Electrochemically induced deposition method to prepare γ-MnO_2/multi-walled carbon nanotube composites as electrode material in supercapacitors. *Mater. Res. Bull.* **2008**, *4*, 2085–2091. [CrossRef]
33. Hiramatsu, M.; Shiji, K.; Amano, H.; Hori, M. Fabrication of vertically aligned carbon nanowalls using capacitively coupled plasma-enhanced chemical vapor deposition assisted by hydrogen radical injection. *Appl. Phys. Lett.* **2004**, *84*, 4708–4710. [CrossRef]
34. Ni, Z.; Wang, Y.; Yu, T.; Shen, Z. Raman spectroscopy and imaging of graphene. *Nano Res.* **2008**, *1*, 273–291. [CrossRef]
35. Masarapu, C.; Zeng, H.F.; Hung, K.H.; Wei, B. Effect of temperature on the capacitance of carbon nanotube supercapacitors. *ACS Nano* **2009**, *3*, 2199–2206. [CrossRef] [PubMed]
36. Pandit, B.; Dubai, D.P.; Gómez-Romero, P.; Kale, B.B.; Sankapal, B.R. V_2O_5 encapsulated MWCNTs in 2D surface architecture: Complete solid-state bendable highly stabilized energy efficient supercapacitor device. *Sci. Rep.* **2017**, *7*, 43430. [CrossRef]
37. Qu, D. Application of a.c. impedance technique to the study of the proton diffusion process in the porous MnO_2 electrode. *Electrochim. Acta* **2003**, *48*, 1675–1684. [CrossRef]
38. Wang, X.; Hao, H.; Liu, J.; Huang, T.; Yu, A. A novel method for preparation of macroporous lithium nickel manganese oxygen as cathode material for lithium ion batteries. *Electrochim. Acta* **2011**, *56*, 4065–4069. [CrossRef]
39. Ouksel, L.; Kerkour, R.; Chelali, N.E. Proton diffusion process manganese dioxide for use in rechargeable alkaline zinc manganese dioxide batteries and its electrochemical performance. *Ionics* **2016**, *22*, 1751–1757. [CrossRef]
40. Li, T.; Beidaghi, M.; Xiao, X.; Huang, L.; Hu, Z.; Sun, W.; Chen, X.; Gogotsi, Y.; Zhou, J. Ethanol reduced molybdenum trioxide for Li-ion capacitors. *Nano Energy* **2016**, *26*, 100–107. [CrossRef]
41. Wei, J.; Nagarajan, N.; Zhitomirsky, I. Manganese oxide films for electrochemical supercapacitors. *J. Mater. Proc. Technol.* **2007**, *186*, 356–361. [CrossRef]
42. Devaraj, S.; Munichandraiah, N. Electrochemical supercapacitor studies of nanostructured α-MnO_2 synthesized by microemulsion method and the effect of annealing. *J. Electrochem. Soc.* **2007**, *154*, A80–A88. [CrossRef]
43. Chang, J.-K.; Chen, Y.-L.; Tsai, W.-T. Effect of heat treatment on material characteristics and pseudo-capacitive properties of manganese oxide prepared by anodic deposition. *J. Power Sources* **2004**, *135*, 344–353. [CrossRef]
44. Lin, C.-K.; Chuang, K.-H.; Lin, C.-Y.; Tsay, C.-Y.; Chen, C.-Y. Manganese oxide films prepared by sol-gel process for supercapacitor application. *Surf. Coat. Technol.* **2007**, *2002*, 1272–1276. [CrossRef]
45. Lang, X.; Hirata, A.; Fujita, T.; Chen, M. Nanoporous metal/oxide hybrid electrodes for electrochemical supercapacitors. *Nat. Nanotechnol.* **2011**, *6*, 232–236. [CrossRef] [PubMed]
46. Toupin, M.; Brousse, T.; Bélanger, D. Charge storage mechanism of MnO_2 electrode used in aqueous electrochemical capacitor. *Chem. Mater.* **2004**, *16*, 3184–3190. [CrossRef]
47. Chang, J.K.; Huang, C.H.; Lee, M.T.; Tsai, W.T.; Deng, M.J.; Sun, I.W. Physicochemical factors that affect the pseudocapacitance and cyclic stability of Mn oxide electrodes. *Electrochim. Acta* **2009**, *54*, 3278–3284. [CrossRef]
48. Tsang, C.; Kim, J.; Manthiram, A. Synthesis of manganese oxides by reduction of $KMnO_4$ with KBH_4 in aqueous solutions. *J. Solid State Chem.* **1998**, *137*, 28–32. [CrossRef]

© 2019 by the authors. Licensee MDPI, Basel, Switzerland. This article is an open access article distributed under the terms and conditions of the Creative Commons Attribution (CC BY) license (http://creativecommons.org/licenses/by/4.0/).

Article

Development of ZIF-Derived Nanoporous Carbon and Cobalt Sulfide-Based Electrode Material for Supercapacitor

Rabia Ahmad [1], Naseem Iqbal [1],* and Tayyaba Noor [2]

[1] US–Pakistan Center for Advanced Studies in Energy (USPCAS–E), National University of Sciences and Technology (NUST), Islamabad 44000, Pakistan
[2] School of Chemical and Materials Engineering (SCME), National University of Sciences and Technology (NUST), Islamabad 44000, Pakistan
* Correspondence: naseem@uspcase.nust.edu.pk; Tel.: +92-51-9085-5281

Received: 8 July 2019; Accepted: 14 August 2019; Published: 11 September 2019

Abstract: Zeolitic Imidazolate Framework (ZIF-67) was prepared in two different solvents—water and methanol. Nanoporous carbon was derived from ZIF-67 via pyrolysis in an inert atmosphere. Anion exchange step of sulfidation on the synthesized material has a great influence on the structure and properties. Structural morphology and thermal stability were characterized by X-ray diffraction (XRD), scanning electron microscopy (SEM)/energy dispersive x-ray spectroscopy (EDS), Brunauer-Emmett-Teller (BET), and thermogravimetric (TG) analysis. The electrochemical analysis was evaluated by cyclic voltammetry, chronopotentiometry, and impedance analysis. The as-prepared nanoporous carbon and cobalt sulfide (NPC/CS) electrode material (water) in 2M KOH electrolyte solution exhibit high specific capacitance of 677 F/g. The excellent electrochemical performance of the NPC/CS was attributed to its hierarchical structure. This functionalized ZIF driven strategy paves the way to the preparation of various metal oxide and metal sulfide-based nanoheterostructures by varying the type of metal.

Keywords: ZIF-67; water; methanol; sulfidation; specific capacitance

---

1. Introduction

With the affluence of the energy industry, several countries have paid much attention to the advancement of power sources. Presently, studies on supercapacitors are rare, compared to the energy storage devices of batteries. As an alternative energy source, supercapacitors have the advantages of high power density, long cyclic stability, and low-cost [1–3]. The main four parts of a supercapacitor are electrode material, electrolyte, separator, and collector. Electrode material exhibits a primary role in the performance of the supercapacitors [4,5].

Among the reported electrode materials for the supercapacitors, metallic oxides and metallic hydroxides have very good theoretical specific capacity but they have very poor electrical conductivity and poor cyclic stability [6]. At present, metallic sulfides have gained many attractions for the applications in supercapacitors as they have very good specific capacitance and electrical conductivity, supporting the improved electrochemical features [7]. Metalic sulfides electrode materials have a high reversible Faradic reaction at the electrode/electrolyte interface that occurs during the charge transfer process. A number of sulfides use supercapacitor applications, i.e., MnS [8], CuS [9], MoS [10], CoS [11], NiS [12], as well as bimetallic sulfides such as $NiCo_2S_4$ [7]; although, the enhancement in surface area, porosity, and mechanical support can improve the electrochemical properties of the metallic sulfides. For that purpose, graphene/graphene oxide and metallic sulfide-based composed like NiS/reduced GO,

Co₃S₄ growth on graphene have been reported as a promising electrode material for the applications in supercapacitors [11].

The metal-organic frameworks (MOFs) have deployed a great sway in the development of supercapacitors since the MOFs were formed in the late 1990s [13]. Excessive consideration was paid to ZIF-67 just because of its polyhedral framework. Due to the number of porosities, considerable surface area, small density, thermal, and chemical stabilities [1], ZIFs represent a breakthrough in the various applications comprising adsorption/separation [14,15], sensors [16], catalysis [17], gas storage [18], and drug delivery [19].

Metal-organic framework (MOF) (and by extension, ZIFs) is a class of nanoporous material that is assembled by coordinated bonds between the two main components—metal ions and organic linker—to shape a 3D porous assembly [4]. Tremendous porosity and surface area, exceptional pore size, and chemical permanency, in ZIFs like ZIF-8 and ZIF-67, are extensively useful in numerous applications such as gas storage [2], separation technologies [14], catalysis [16], and energy-related fields [1]. ZIFs with particle-like morphology remained dominant in use so far, and it has been an eye-catching task to control their framework. It is commonly believed that different structural evaluation stages, like the first formation of nucleation, initiation crystallization, and then growth, are tangled in the crystallization of ZIFs [2]. In recent times, a new 2D leaf-like structure was prepared by using metal ions of zinc and the linker 2–methylimidazole. It is well acknowledged that morphology and particle size have a great effect on both the extensive and intensive characteristics of the material [20,21]. ZIFs derived nanoporous carbon (NPC) and metal oxide (MO) based material has a synergistic effect of both carbon-based material and metallic oxide. Furthermore, in comparison with the metal oxide-based material, metal sulfide-based electrode materials have superior electrocatalytic activity [22,23].

In the present work, we synthesized cobalt sulfide onto the nanoporous carbon to positively incorporate the synergistic effect towards the electrical conductivity and stability in two different solvents. Here we demonstrated, the 2D leaf-like morphology exhibits the superior capacity and the best electrochemical performance. Effect of solvent i.e., water and methanol, on the synthesis of ZIF-67-derived nanoporous carbon and cobalt sulfide-based electrode, and then measured specific capacitance, is also investigated for the application in supercapacitor.

2. Materials and Methods

2.1. Chemicals

Metal ion used is cobalt nitrate hexahydrate (Co(NO₃)₂·6H₂O, 99%), and the linker is 2–Methylimidazole (99%) were used. All the chemicals were purchased from Sigma Aldrich/Merck, have analytical purity and used as received.

2.2. Synthesis of ZIF-67

For the synthesis of ZIF-67, the following scheme was used; 0.873 g of cobalt nitrate hexahydrate was dissolved in 30 mL of methanol to form a clear solution; 0.984 g of organic linker 2–Methylimidazole was dissolved in 10 mL of methanol to make another clear solution. The two solutions mixed with a vigorous shake of a few minutes. The mixed solution was kept overnight at room temperature. Thenceforth, centrifugation is used to collect the precipitates followed by multiple washes using methanol and dried up at 80 °C for 6 hours. The same experimental procedure was used to synthesis ZIF-67 using deionized water.

2.3. Preparation of Nanoporous Carbon (NPC) and Cobalt Oxide (CO)

The dried powder ZIF-67 particles were heated at 350 °C for 1.5 hours, raised to 700 °C at a ramp rate of 4 °C per minute, followed by pyrolysis for 3.5 hours under a flowing argon atmosphere. Next, the prepared black fluffy powder was cooled to room temperature naturally.

2.4. Preparation of Nanoporous Carbon (NPC) and Cobalt Sulfide (CS)

Aqueous suspension of nanoporous carbon-containing cobalt oxide was stirred in 0.015M sodium sulfide for 30 minutes, and then solution mixture transferred to a stainless steel autoclave of Teflon-lined and heated at 120 °C for 6 hours. As obtained precipitates washed and dried.

2.5. Material Characterization

The surface morphological analysis of the polyhedral structure was described by field emission scanning electron microscopy (VEGA3, 51–ADD0007) (Tescan, Brno, Czech Republic). The identification of a crystalline structure was elucidated by X-ray diffraction on a diffractometer (D8 Advance, CuKR, λ = 1.54Å) (Bruker, Karlsruhe, Germany). The thermogravimetric analysis was conducted on a DTG–60H (Shimadzu, Kyoto, Japan) instrument in the temperature range of room temperature to 800 °C. The surface areas and porous structure of synthesized NPC/CO and NPC/CS were measured by Brunauer–Emmett–Teller analysis using NovaWin 20e (Quantachrome, Virginia, USA) instrument at a relative pressure p/p^o = 0–1.0 and the samples were degassed at 160 °C under the vacuum.

2.6. Electrochemical Testing

The electrochemical measurements of the prepared samples were performed on an electrochemical workstation CHI 760E (CH Instrument, Texas, USA) with a setup of three electrodes. A reference electrode Ag/AgCl (SC) and a counter electrode of a platinum coil are used. To prepare the ink for the working electrode (GC), the following scheme was used; catalyst (2 mg) was dispersed ultrasonically for 1 to 2 hours in 0.08 mL of ethanol solution with 0.02 mL of Nafion solution (5 wt. %) to form a homogeneous ink. Then, the polished glassy carbon electrode (3 mm diameter) was coated by dropping the suspension (5 µL).

Electrochemical impedance spectroscopy (EIS) used a frequency field of 1 to 100 kHz in 2 M KOH solution. Cyclic voltammetry was performed within the potential window of 0.5 V in 2 M KOH solution with various sweep rates of 10, 20, 50, 80, and 100 mV s^{-1}. The chronopotentiometry technique is used to measure the charge-discharge curve at 0.01 mA cathodic current to obtain the discharge time in the potential window of 0.35 V.

3. Results

3.1. Morphology/Structural Analysis

The framework and surface morphology of the as-prepared specimen were explored on SEM (Figure 1). Metal-organic framework (MOF) ZIF-67 displayed a different framework of structure with a different solvent. ZIF-67 prepared in methanol exhibits well-defined polyhedrons with a smooth surface. The shape of ZIF-67 showed typical and uniform rhombic dodecahedron assembly, which is consistent with the morphology of ZIF-67 in the literature [4]. The synthesized product with water as a solvent (Figure 1a) showed 2D leaf-like morphology, and the side length of this leaf-like morphology is approximately 2.498 micron. In Figure 1d, in the rhombic face, the side length is ~430 nm. The appearance of each rhombic face is smooth, proposing high purity of the as-prepared product ZIF-67. Furthermore, the sharp edges and clear-cut corners of ZIF-67 particles (see Figure 1d) demonstrate the ascertaining of the crystallography characteristics.

The crystal structure of ZIF-67 before and after pyrolysis was examined by XRD measurements. The relative intensity and peak positions are well-matched with the literature [24], which shows that ZIF-67 was successfully synthesized. The wide diffraction peak at 2θ =25° belongs to peak (002) of graphite carbon. The diffraction peaks derived for the metallic cobalt at the (111) phase and (200) phase were detected at 2θ= 45° and 52° (JCPDS card No. 15–0806). All diffraction peaks of nanoporous carbon and cobalt sulfide, as shown in Figure 2, correspond well with the patterns reported in the previous study [25,26]. Through the sulfidation treatment, CoO particles were successfully transformed into the well-defined CoS$_2$ cubic phase (JCPDS card No. 41–1471).

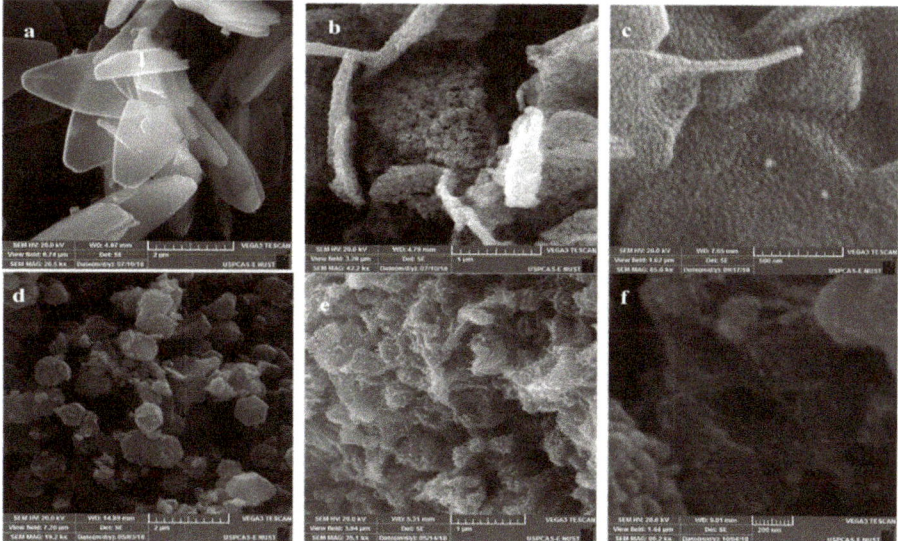

Figure 1. SEM images of ZIF-67 prepared in different solvents: (**a**) ZIF-67 prepared in H$_2$O, before pyrolysis, after pyrolysis; (**b**) NPC/CO (H$_2$O) at 700 °C; (**c**) NPC/CS (H$_2$O); (**d**) ZIF-67 prepared in MeOH; (**e**) NPC/CO (MeOH) at 700 °C; and (**f**) NPC/CS (MeOH).

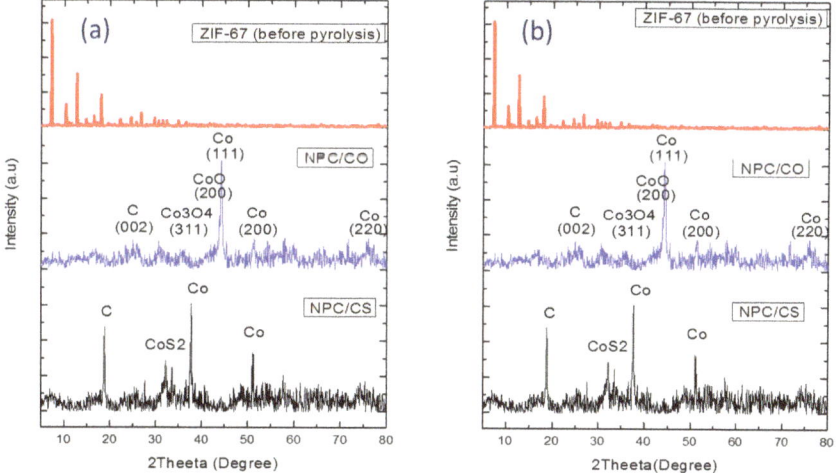

Figure 2. XRD patterns of (**a**) ZIF-67, NPC/CO and NPC/CS prepared in H2O and (**b**) ZIF-67, NPC/CO, and NPC/CS prepared in MeOH.

The atomic ratio of C: Co:O:S is confirmed by the energy-dispersive X-ray spectroscopy (EDS) analyses (Figure 3).

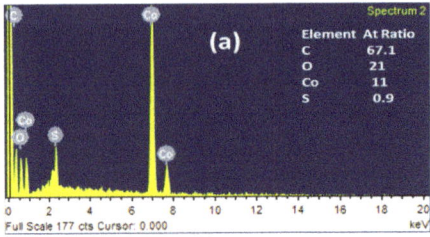
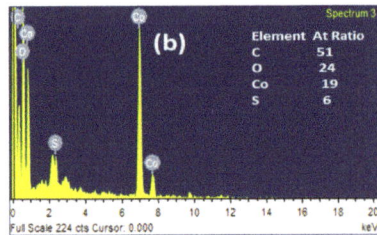

Figure 3. EDS spectrum of (**a**) NPC/CS (H2O) and (**b**) NPC/CS (MeOH).

TGA curves of prepared ZIF-67, nanoporous carbon/cobalt oxide (NPC/CO), and nanoporous carbon/cobalt sulfide (NPC/CS) under flowing nitrogen condition, as shown in Figure 4. The first stage mass loss of approximately 11%, below 200 °C, and a 6% loss from 200 °C to 320 °C correspond to the removal of guest water molecules, i.e., surface moisture, solvent, nitrates, and weekly bounded linker molecules, respectively [27]. Next, the rapid degradation occurred with the mass losses from 320 °C in H_2O-based system and 316°C MeOH-based system. Finally, no further weight loss was observed after 530°C of NPC/CS (H2O) and NPC/CS (MeOH), the quasistatic point, and the residual mass was comprised of metal oxide [28].

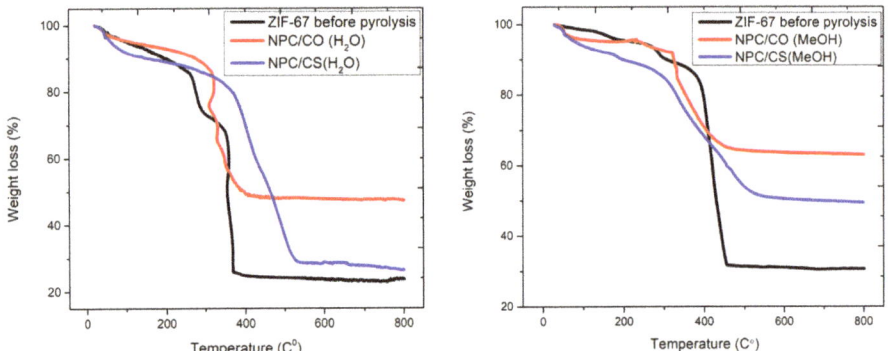

Figure 4. TGA curve analysis of as prepared ZIF-67, NPC/CO, and NPC/CS.

In the thermal analysis, Figure 4 curves indicate that nanoporous carbon and cobalt sulfide doped samples have high thermal stability (Table 1).

Table 1. TGA thermal analysis of ZIF-derived nanoporous carbon and cobalt oxide/sulfide-based materials.

Compound	Temp. Range (°C)	Mass Loss or Residue (%)
ZIF-67 (MeOH) before pyrolysis	336–458	57
ZIF-67 (H_2O) before pyrolysis	302–373	48
NPC/CO (MeOH)	314–492	28
NPC/CO (H_2O)	297–414	40
NPC/CS (MeOH)	300–544	34
NPC/CS (H_2O)	311–535	56

The N_2 adsorption-desorption isotherms for as-prepared samples, which is characterized as type I and type II hysteresis loops according to the IUPAC classification, indicating the microporous characteristics of the synthesized sample. Nitrogen adsorption can be considered as the first stage in the characterization of microporous and mesoporous solids. In the vision of the complexity of the condensation and evaporation mechanisms, one should not be supposed to be able to conclude a reliable pore size distribution unless certain conditions are met. It is recommended that the shape and location of the hysteresis loop should always be taken into account before any computation [29].

It can be identified that the specific surface areas of NPC/CS have increased largely with the treatment of sulfidation, as the analysis is presented in Table 2. From the Figure 5, it can be clearly seen Type I isotherm is given by all the materials except NPC/CS (MeOH). Type I isotherms are microporous materials that acquire mostly wider micropores and possibly narrow mesopores. In the case of NPC/CS (MeOH), type II isotherm materials are often disordered and the distribution of pore size and shape is not well-defined [30]. The type II isotherm is the consequence of the open monolayer-to-multilayer adsorption up to p/p_0.

Table 2. Surface area, pore volume, and average pore size of NPC/CO and NPC/CS.

Electrode Material	Surface Area (m²/g)	Pore Volume (cm³/g)	Average Pore Size (nm)
NPC/CO (MeOH)	726.3	0.273	1.277
NPC/CS (MeOH)	934.5	0.532	1.677
NPC/CO (H₂O)	264.6	0.063	1.544
NPC/CS (H₂O)	521.4	0.141	2.25

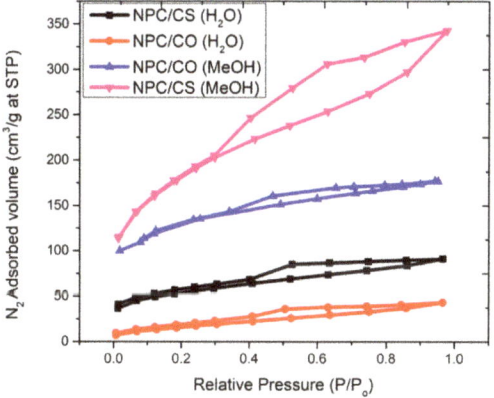

Figure 5. N$_2$ adsorption isotherms at 77 K.

The enhanced surface area can be attributed to smaller diameters and larger quantities of the nanocrystals and nanopores. The total pore volume of NPC/CS showed an increment as compared to parent NPC/CO.

3.2. Electrochemical Testing

The measurement and investigation of electrochemical behavior were explored by galvanostatic charge-discharge (GCD), cyclic voltammetry (CV), and electrochemical impedance spectroscopy (EIS) in 2M KOH solution.

By definition, capacitance = charge/voltage [31].

Capacitance value can be calculated from resulting cyclic voltammogram using the following equation [32].

$$C = \frac{Q}{2\Delta V m} \tag{1}$$

where C is the specific capacitance in (Figure 1), Q is the average integral area under the curve, ΔV is the potential window in volts, and m is the mass loading of the active material in the working electrode (g).

The energy density (ED) uses the relation [32]:

$$ED = \frac{1}{2} \cdot C \frac{(V)^2}{3.6} \qquad (2)$$

where C is the value of specific capacitance and V is the voltage window.

The reversible process was observed in all cases. The linear trend across the whole range of scan rates reveals that the process is reversible.

The electrochemical performance of ZIF-67-derived NPC/CO and NPC/CS were studied by varying the scan rate ranging from 10mV/s to 100 mV/s CV curves and shown in Figure 6. The anodic peaks and cathodic peaks, which are related to positive current and negative current, respectively, in the CV curves, originate from the oxidation and reduction process of the cobalt cation, which indicates that the capacitance aspects are primarily driven by Faradaic redox reaction.

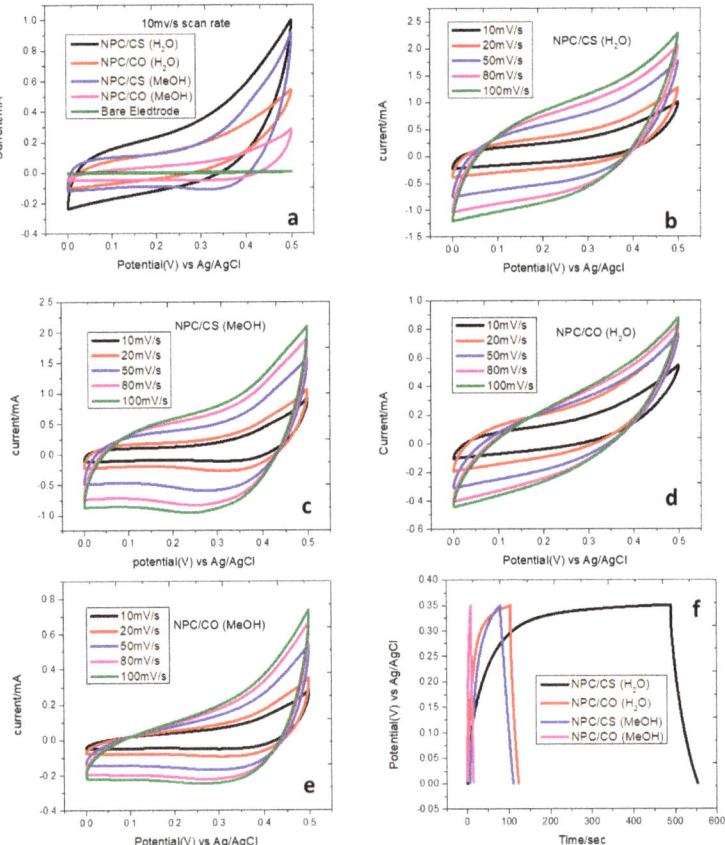

Figure 6. Electrochemical properties of the NPC/CO and NPC/CS in 2M KOH: (**a**) cyclic voltammetry at 10 mv/s scan rate (**b–e**) at different scan rates and (**f**) charge-discharge profiles at 0.01 mA/g.

It was observed that as the scan rate increases from 10mV/s to 100mV/s, the specific capacitance increases. This variation in capacitance reveals that at low scan rate, inner parts and outer part of the nanoporous material exhibited the redox reaction, whereas, for high scan rate, the only outer part of the material involved redox reaction [33,34]. There are interfaces faces present in the case of oxides that causes the limitation in the connectivity or the flow of electrons, while this is reported by many of authors that the deposition of sulfide on oxide containing composite reduces these interfaces and produce connectivity in the flow. This will contribute to enhancing conductivity [11,35]. Thus the specific capacitance and the energy density calculated is given in Table 3.

Table 3. Specific capacitance calculated from cyclic voltammetry.

Electrode Material	Specific Capacitance (F/g)	Energy Density (Wh/kg)
NPC/CO (MeOH)	159	5.520
NPC/CS (MeOH)	480	16.666
NPC/CS (H_2O)	373	12.951
NPC/CS (H_2O)	677	23.506

3.3. Cyclic Stability Study

One of the most essential and considerable parts in the achievement of a supercapacitor is cyclic stability. Figure 7 shows the electrochemical performance of the nanoporous carbon; the cobalt sulfide-based electrode recorded over 1600 cycles with a scan rate of 100 mV/s. The cyclic stability study demonstrates the change in the specific capacitance of the NPC/CS electrode with the number of cycles. The specific capacitance decreases with 86% retention capacity over 1600 cycles for the case of NPC/CS (H_2O) and noted as 74% over 1600 cycles for the NPC/CS (MeOH) systems. Zhu et al. [36] presented nearly 74% stability aimed at the nickel sulfide electrode material by recording 1000 cycles in a potassium hydroxide electrolyte; a nearly identical method to that of the cinnamon-like electrode. This decomposition was perceived in cycling stability. This might be expected because of the disintegration of the active electrode material in the electrolyte and the capacity imbalances between the electrochemical electrodes, which originates the instability of the electrode potential [37].

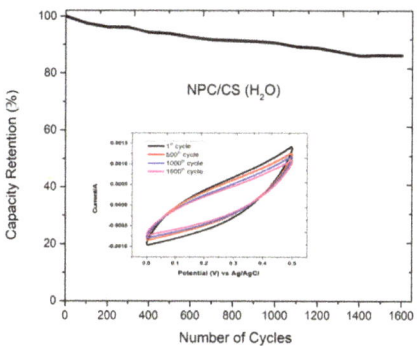
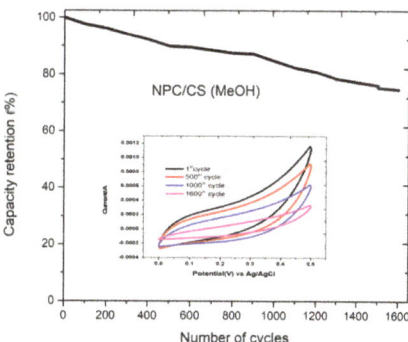

Figure 7. Numbers of cycles versus capacitive retention for the electrode NPC/CS electrode materials recorded over 1600 cycles at a scan rate of 100 mV/s.

3.4. Electrochemical Impedance Spectroscopy Study

To evaluate the frequency performance and confrontation attitude of any material, electrochemical impedance spectroscopy is a robust engine [38]. The EIS spectra are recorded in a frequency ranging from 100 kHz to 1 Hz and shown in Figure 8. The perfect supercapacitor has a greater slop in a low-frequency region, which signifies electro–capacitive behavior [39,40].

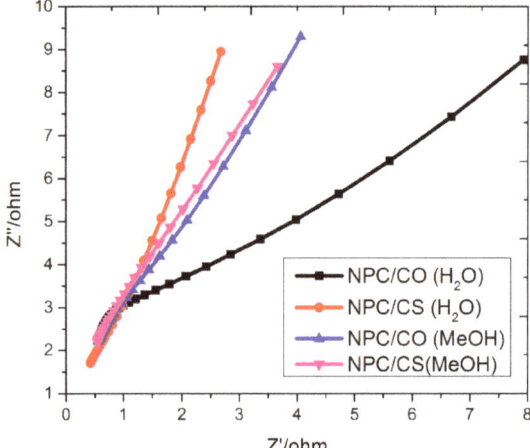

Figure 8. Nyquist plot of the NPC/CO and NPC/CS.

It can be seen from Figure 8 that the smaller arc diameter in the EIS spectrum of nanoporous carbon and cobalt sulfide-based materials showed lower charge transfer resistance and more idealistic properties as compared to nanoporous carbon and metal oxide electrodes [41]. The slop of NPC/CS (water) approach has ideally straight line, inferring the superior accessibility of ions.

4. Conclusions

In summary, we have successfully fabricated ZIF-67-derived nanoporous carbon and metal sulfide-based electrode material in a simple and economical method. SEM and XRD confirmed the synthesis of nanoporous material with increased surface area and 2D morphology. The enhanced electrochemical performance, due to sulfidation of cobalt oxide, was investigated by CV GCD and EIS. The specific capacitance value has increased to 677 F/g in the case of NPC/CS (H_2O). A novel and facile route for the synthesis of nanoporous and binder-free electrode material is proposed with increased specific capacitance for the energy storage application.

Author Contributions: Conceptualization, R.A. and N.I.; data curation, methodology, and investigation, R.A.; Supervision N.I. and T.N., original draft preparation R.A.; writing-reviews and editing all authors.

Funding: Funding for this project was provided by USAID and U.S.-Pakistan Centers for Advanced Studies in Energy (USPCAS-E), NUST, Pakistan.

Acknowledgments: The authors are very grateful to USAID and U.S.-Pakistan Centers for Advanced Studies in Energy, NUST, Pakistan.

Conflicts of Interest: The authors declare no conflict of interest.

References

1. Chen, B.; Yang, Z.; Zhu, Y.; Xia, Y. Zeolitic imidazolate framework materials: Recent progress in synthesis and applications. *J. Mater. Chem. A* **2014**, *2*, 16811–16831. [CrossRef]
2. Yao, J.; Wang, H. Zeolitic imidazolate framework composite membranes and thin films: Synthesis and applications. *Chem. Soc. Rev.* **2014**, *43*, 4470–4493. [CrossRef] [PubMed]
3. Park, K.S.; Ni, Z.; Côté, A.P.; Choi, J.Y.; Huang, R.; Uribe–Romo, F.J.; Chae, H.K.; O'Keeffe, M.; Yaghi, O.M. Exceptional chemical and thermal stability of zeolitic imidazolate frameworks. *P. Natl. Acad. Sci. USA* **2006**, *103*, 10186–10191. [CrossRef] [PubMed]

4. Wang, Q.; Gao, F.; Xu, B.; Cai, F.; Zhan, F.; Gao, F.; Wang, Q. ZIF–67 derived amorphous CoNi2S4 nanocages with nanosheet arrays on the shell for a high–performance asymmetric supercapacitor. *Chem. Eng. J.* **2017**, *327*, 387–396. [CrossRef]
5. Zhou, J.J.; Han, X.; Tao, K.; Li, Q.; Li, Y.L.; Chen, C.; Han, L. Shish–kebab type MnCo2O4@Co3O4 nanoneedle arrays derived from MnCo–LDH@ZIF–67 for high–performance supercapacitors and efficient oxygen evolution reaction. *Chem. Eng. J.* **2018**, *354*, 875–884. [CrossRef]
6. Guan, C.; Li, X.; Wang, Z.; Cao, X.; Soci, C.; Zhang, H.; Fan, H.J. Nanoporous walls on macroporous foam: Rational design of electrodes to push areal pseudocapacitance. *Adv. Mater.* **2012**, *24*, 4186–4190. [CrossRef]
7. Hao, P.; Tian, J.; Sang, Y.; Tuan, C.C.; Cui, G.; Shi, X.; Wong, C.P.; Tang, B.; Liu, H. 1D Ni–Co oxide and sulfide nanoarray/carbon aerogel hybrid nanostructures for asymmetric supercapacitors with high energy density and excellent cycling stability. *Nanoscale* **2016**, *8*, 16292–16301. [CrossRef]
8. Tang, Y.; Chen, T.; Yu, S. Morphology controlled synthesis of monodispersed manganese sulfide nanocrystals and their primary application in supercapacitors with high performances. *Chem. Commun.* **2015**, *51*, 9018–9021. [CrossRef]
9. Raj, C.J.; Kim, B.C.; Cho, W.J.; Lee, W.G.; Seo, Y.; Yu, K.H. Electrochemical capacitor behavior of copper sulfide (CuS) nanoplatelets. *J. Alloy. Compd.* **2014**, *586*, 191–196.
10. Huang, K.J.; Wang, L.; Liu, Y.J.; Liu, Y.M.; Wang, H.B.; Gan, T.; Wang, L.L. Layered MoS2–graphene composites for supercapacitor applications with enhanced capacitive performance. *Int. J. Hydrog. Energy* **2013**, *38*, 14027–14034. [CrossRef]
11. Sun, L.; Lu, L.; Bai, Y.; Sun, K. Three–dimensional porous reduced graphene oxide/sphere–like CoS hierarchical architecture composite as efficient counter electrodes for dye–sensitized solar cells. *J. Alloy. Compd.* **2016**, *654*, 196–201. [CrossRef]
12. Yang, J.; Duan, X.; Guo, W.; Li, D.; Zhang, H.; Zheng, W. Electrochemical performances investigation of NiS/rGO composite as electrode material for supercapacitors. *Nano Energy* **2014**, *5*, 74–81. [CrossRef]
13. Zhou, H.C.; Long, J.R.; Yaghi, O.M. Introduction to Metal–Organic Frameworks. *Chem. Rev.* **2012**, *112*, 673–674. [CrossRef] [PubMed]
14. Bux, H.; Chmelik, C.; Krishna, R.; Caro, J. Ethene/ethane separation by the MOF membrane ZIF-8: Molecular correlation of permeation, adsorption, diffusion. *J. Membrane. Sci.* **2011**, *369*, 284–289. [CrossRef]
15. Song, X.D.; Wang, S.; Hao, C.; Qiu, J.S. Investigation of SO2 gas adsorption in metal–organic frameworks by molecular simulation. *Inorg. Chem. Commun.* **2014**, *46*, 277–281. [CrossRef]
16. Samadi-Maybodi, A.; Ghasemi, S.; Ghaffari-Rad, H. A novel sensor based on Ag–loaded zeolitic imidazolate framework-8 nanocrystals for efficient electrocatalytic oxidation and trace level detection of hydrazine. *Sens. Actuators B Chem.* **2015**, *220*, 627–633. [CrossRef]
17. Guan, Y.; Shi, J.; Xia, M.; Zhang, J.; Pang, Z.; Marchetti, A.; Wang, X.; Cai, J.; Kong, X. Monodispersed ZIF-8 particles with enhanced performance for CO2 adsorption and heterogeneous catalysis. *Appl Surf. Sci* **2017**, *423*, 349–353. [CrossRef]
18. Mu, L.; Liu, B.; Liu, H.; Yang, Y.; Sun, C.; Chen, G. A novel method to improve the gas storage capacity of ZIF-8. *J. Mater. Chem.* **2012**, *22*, 12246–12252. [CrossRef]
19. Shearier, E.; Cheng, P.; Zhu, Z.; Bao, J.; Hu, Y.H.; Zhao, F. Surface defection reduces cytotoxicity of Zn(2–methylimidazole)2 (ZIF-8) without compromising its drug delivery capacity. *Rsc Adv.* **2016**, *6*, 4128–4135. [CrossRef]
20. Wang, Q.; Bai, J.; Lu, Z.; Pan, Y.; You, X. Finely tuning MOFs towards high–performance post–combustion CO2 capture materials. *Chem. Commun.* **2016**, *52*, 443–452. [CrossRef]
21. Bustamante, E.L.; Fernández, J.L.; Zamaro, J.M. Influence of the solvent in the synthesis of zeolitic imidazolate framework-8 (ZIF-8) nanocrystals at room temperature. *J. Colloid Interface Sci.* **2014**, *424*, 37–43. [CrossRef] [PubMed]
22. Yang, C.; Ou, X.; Xiong, X.; Zheng, F.; Hu, R.; Chen, Y.; Liu, M.; Huang, K. V5S8–graphite hybrid nanosheets as a high rate–capacity and stable anode material for sodium–ion batteries. *Energy Environ. Sci.* **2017**, *10*, 107–113. [CrossRef]
23. Zhu, X.; Liang, X.; Fan, X.; Su, X. Fabrication of flower–like MoS2/TiO2 hybrid as an anode material for lithium ion batteries. *Rsc. Adv.* **2017**, *7*, 38119–38124. [CrossRef]

24. Wu, R.; Qian, X.; Rui, X.; Liu, H.; Yadian, B.; Zhou, K.; Wei, J.; Yan, Q.; Feng, X.Q.; Long, Y.; et al. Zeolitic Imidazolate Framework 67–Derived High Symmetric Porous Co3O4 Hollow Dodecahedra with Highly Enhanced Lithium Storage Capability. *Small* **2014**, *10*, 1932–1938. [CrossRef] [PubMed]
25. Torad, N.L.; Hu, M.; Ishihara, S.; Sukegawa, H.; Belik, A.A.; Imura, M.; Ariga, K.; Sakka, Y.; Yamauchi, Y. Direct Synthesis of MOF–Derived Nanoporous Carbon with Magnetic Co Nanoparticles toward Efficient Water Treatment. *Small* **2014**, *10*, 2096–2107. [CrossRef]
26. Hu, H.; Han, L.; Yu, M.; Wang, Z.; Lou, X.W. Metal–organic–framework–engaged formation of Co nanoparticle–embedded carbon@Co9S8 double–shelled nanocages for efficient oxygen reduction. *Energy Environ. Sci.* **2016**, *9*, 107–111. [CrossRef]
27. Escorihuela, J.; Sahuquillo, Ó.; García–Bernabé, A.; Giménez, E.; Compañ, V. phosphoric acid doped polybenzimidazole (PBI)/zeolitic imidazolate framework composite membranes with significantly enhanced proton conductivity under low humidity conditions. *Nanomaterials* **2018**, *8*, 775. [CrossRef]
28. Barjola, A.; Escorihuela, J.; Andrio, A.; Giménez, E.; Compañ, V. Enhanced conductivity of composite *membranes* based on sulfonated poly (ether ether ketone)(SPEEK) with zeolitic imidazolate frameworks (ZIFs). *Nanomaterials* **2018**, *8*, 1042. [CrossRef]
29. Qin, J.; Wang, S.; Wang, X. Visible–light reduction CO2 with dodecahedral zeolitic imidazolate framework ZIF–67 as an efficient co–catalyst. *Appl. Catal. B: Environ* **2017**, *209*, 476–482. [CrossRef]
30. Alothman, Z. A Review: Fundamental Aspects of Silicate Mesoporous Materials. *Materials* **2012**, *5*, 2874–2902. [CrossRef]
31. Chen, G.Z. Understanding supercapacitors based on nano–hybrid materials with interfacial conjugation. *Prog. Nat. Sci–Mater* **2013**, *2*, 245–255. [CrossRef]
32. Ensafi, A.A.; Ahmadi, N.; Rezaei, B. Electrochemical preparation and characterization of a polypyrrole/nickel–cobalt hexacyanoferrate nanocomposite for supercapacitor applications. *Rsc. Adv.* **2015**, *5*, 91448–91456. [CrossRef]
33. Dong, X.; Wang, L.; Wang, D.; Li, C.; Jin, J. Layer–by–Layer Engineered Co–Al Hydroxide Nanosheets/Graphene Multilayer Films as Flexible Electrode for Supercapacitor. *Langmuir.* **2012**, *28*, 293–298. [CrossRef] [PubMed]
34. Zhou, Y.; Liu, C.; Li, X.; Sun, L.; Wu, D.; Li, J.; Huo, P.; Wang, H. Chemical precipitation synthesis of porous Ni2P2O7 nanowires for supercapacitor. *J. Alloy. Compd.* **2019**, *790*, 36–41. [CrossRef]
35. Syed, J.A.; Ma, J.; Zhu, B.; Tang, S.; Meng, X. Hierarchical Multicomponent Electrode with Interlaced Ni(OH)$_2$ Nanoflakes Wrapped Zinc Cobalt Sulfide Nanotube Arrays for Sustainable High-Performance Supercapacitors. *Adv. Energy Mater.* **2017**, *7*, 1701228. [CrossRef]
36. Zhu, B.T.; Wang, Z.; Ding, S.; Chen, J.S.; Lou, X.W. Hierarchical nickel sulfide hollow spheres for high performance supercapacitors. *Rsc. Adv.* **2011**, *1*, 397–400. [CrossRef]
37. Zhu, T.; Koo, E.R.; Ho, G.W. Shaped–controlled synthesis of porous NiCo2O4 with 1–3 dimensional hierarchical nanostructures for high–performance supercapacitors. *Rsc. Adv.* **2015**, *5*, 1697–1704. [CrossRef]
38. Li, J.; Ren, Z.; Ren, Y.; Zhao, L.; Wang, S.; Yu, J. Activated carbon with micrometer–scale channels prepared from luffa sponge fibers and their application for supercapacitors. *Rsc. Adv.* **2014**, *4*, 35789–35796. [CrossRef]
39. Yang, J.; Gunasekaran, S. Electrochemically reduced graphene oxide sheets for use in high performance supercapacitors. *Carbon* **2013**, *51*, 36–44. [CrossRef]
40. Wang, R.; Li, Q.; Cheng, L.; Li, H.; Wang, B.; Zhao, X.S.; Guo, P. Electrochemical properties of manganese ferrite–based supercapacitors in aqueous electrolyte: The effect of ionic radius. *Colloids Surf. A Phys. Eng. Asp.* **2014**, *457*, 94–99. [CrossRef]
41. Sun, K.; Feng, E.; Peng, H.; Ma, G.; Wu, Y.; Wang, H.; Lei, Z. A simple and high–performance supercapacitor based on nitrogen–doped porous carbon in redox–mediated sodium molybdate electrolyte. *Electrochim. Acta* **2015**, *158*, 361–367. [CrossRef]

© 2019 by the authors. Licensee MDPI, Basel, Switzerland. This article is an open access article distributed under the terms and conditions of the Creative Commons Attribution (CC BY) license (http://creativecommons.org/licenses/by/4.0/).

Review

AC-Filtering Supercapacitors Based on Edge Oriented Vertical Graphene and Cross-Linked Carbon Nanofiber

Wenyue Li [1], Nazifah Islam [2], Guofeng Ren [1], Shiqi Li [3] and Zhaoyang Fan [1,*]

[1] Department of Electrical and Computer Engineering and Nano Tech Center, Texas Tech University, Lubbock, TX 79409, USA; wenyue.li@ttu.edu (W.L.); renapply@gmail.com (G.R.)
[2] BaoNano, LLC, Lubbock, TX 79415, USA; nzfh.buet@gmail.com
[3] College of Electronic Information, Hangzhou Dianzi University, Hangzhou 310018, China; sqli@hdu.edu.cn
* Correspondence: zhaoyang.fan@ttu.edu; Tel.: +01-806-834-6723

Received: 15 January 2019; Accepted: 14 February 2019; Published: 18 February 2019

Abstract: There is strong interest in developing high-frequency (HF) supercapacitors or electrochemical capacitors (ECs), which can work at the hundreds to kilo hertz range for line-frequency alternating current (AC) filtering in the substitution of bulky aluminum electrolytic capacitors, with broad applications in the power and electronic fields. Although great progress has been achieved in the studies of electrode materials for ECs, most of them are not suitable to work in this high frequency range because of the slow electrochemical processes involved. Edge-oriented vertical graphene (VG) networks on 3D scaffolds have a unique structure that offers straightforward pore configuration, reasonable surface area, and high electronic conductivity, thus allowing the fabrication of HF-ECs. Comparatively, highly conductive freestanding cross-linked carbon nanofibers (CCNFs), derived from bacterial cellulose in a rapid plasma pyrolysis process, can also provide a large surface area but free of rate-limiting micropores, and are another good candidate for HF-ECs. In this mini review, advances in these fields are summarized, with emphasis on our recent contributions in the study of these materials and their electrochemical properties including preliminary demonstrations of HF-ECs for AC line filtering and pulse power storage applications.

Keywords: vertical graphene; cross-linked carbon nanofiber; high-rate supercapacitor; AC filtering; pulse power storage

---

1. Introduction

Of the different dielectric based electrostatic capacitors, aluminum electrolytic capacitors (AECs) have large capacitance densities, commonly used for power related filtering applications such as line-frequency AC filtering, noise decoupling and filtering, direct current (DC) link circuits for variable-frequency drives, pulse power storage and generation, for which, the requirement of frequency response is generally in the hundreds to kilo hertz range. To achieve a large surface area and hence a favorable capacitance density, the aluminum foil electrode is electrochemically etched into a porous structure, which is conformally coated by the electrochemically formed aluminum oxide dielectric layer. A liquid or a polymer electrolyte is then backfilled into the sub-micrometer pores to act as the counter electrode, which is further connected to a current collector, thus constituting an electrostatic capacitor. However, even with an enlarged surface area, the capacitance density of the AEC is still limited, resulting in its bulky size in comparison with the dramatically reduced electronic chips on circuit boards. Great needs exist in downscaling the bulky AEC for compact circuit board design. Considering the much larger capacitance density offered by an electrical double-layer capacitor (EDLC) than the electrostatic capacitor, an interesting question is whether an electrical double-layer based EC, with a compact size, could be used to replace the AEC [1].

Conventional ECs, with their capability of charging and discharging in minutes or seconds, are manufactured for storing or releasing energy burst in supplementing slow batteries or work independently. As explained by the phase angle plot in Figure 1a, an ideal capacitor should have a phase angle close to $-90°$ within its working frequency. The characteristic frequency (f_0) where the phase angle reaches to $-45°$ is defined to delineate the frontier between capacitive and resistive dominance. Compared with conventional ECs having very limited frequency response (<1 Hz), AECs normally show an f_0 up to kHz or even MHz, giving them the capability of being used as a filter capacitor in power systems or electronic circuits. Since the report of ECs with frequency response reaching to the kHz range by using vertical graphene in 2010 by Miller et al. [2], efforts in the investigation of HF-ECs have led to dramatic progress [3–9] in this niche area of supercapacitor technology, as recently reviewed [1]. Different carbon-based nanomaterials including carbon black nanoparticle [10,11], carbon foam [12], carbon nanofiber [13], and carbon nanotubes (CNTs) [14,15], vertical aligned carbon nanotubes [16,17], ordered mesoporous carbon [18,19], graphene foam [3], thin graphene mesh [20] and vertical graphene (VG) [21–24], among others, have been reported as electrodes for HF-ECs. Both sandwich-type and co-planar interdigitated layouts were used in the configuration design of HF-ECs. In addition to the highly conductive aqueous electrolytes, organic and ionic liquid electrolytes [25] as well as polymer electrolytes for solid state capacitors [26], were also studied to boost the rated voltage of the devices. The low-voltage line-frequency AC-filtering function of HF-ECs has been preliminarily demonstrated [18,27], as well as their capability to store pulse energy for environmental energy harvesting [13]. Except for these carbon based electrodes, it deserves to be mentioned that other conductive and electrochemically active materials have also been synthesized and tested for high-rate ECs [28–33], although their speed is still too slow to meet the requirements of HF-ECs at present. For comparison, the structural characteristics and electrochemical performances of these newly studied electrodes are summarized in Table 1.

Table 1. Comparison of different carbon-based materials for HF-ECs in terms of equivalent serial resistance (ESR), phase angle (Φ_{120}), and single electrode areal capacitance (C_A^{120}) at 120 Hz, characteristic frequency (f_0) and type of electrolyte.

Materials	Electrolyte	ESR (Ωcm^2)	Φ_{120} (°)	f_0 (kHz)	C_A^{120} (mF cm^{-2})	Ref.
VG on Ni foil	KOH	0.1	−82	15	0.175	[2]
ErGO foam on Au	KOH	0.14	−84	4.2	0.566	[3]
VG on Ni foam	KOH	0.12	−82	4.0	0.72	[5]
VG on carbonized cellulose paper	KOH	0.04	−83	12 & 5.6	0.6 & 1.5	[6]
PEDOT:PSS on graphene layer	H_2SO_4	0.09	−83.6	>1.0	1.988	[8]
Carbon black on conductive vinyl	KOH	0.39	−75	0.641	1.1	[10]
Laser reduced carbon nanodots	TBAPF$_6$/AN	-	−78	0.955	0.518	[11]
Carbon fiber foam	TEABF$_4$/AN	-	−80.1	2.885	0.264	[12]
Cross-linked carbon nanofibers	KOH	0.009	−82	3.3	2.98	[13]
Ultrathin CNTs film	TEABF$_4$/AN	~ 0.26	−82.2	1.995	0.56	[14]
CNT film	K_2SO_4	0.11	−81	1.425	1.202	[15]
Aligned CNT on graphene film	KOH	0.065	−84.8	1.98	2.72	[16]
Graphitic ordered mesoporous carbon/CNT	TEABF$_4$/AN	0.25	−80.0	>1.0	1.12	[18]
Graphene nano-mesh film	KOH	0.39	−82.3	6.211	0.612	[20]
VG on Ni foil	KOH	0.1	−85.0	20.0	0.53	[21]
VG on carbon fibers	KOH	0.05	−81.5	22.0	0.37	[22]
Carbon black/VG on Ni foil	KOH	0.088	−80	1.0	2.30	[24]

For AC line-frequency filtering applications, the 60 Hz line frequency becomes 120 Hz after a full-bridge rectification, and therefore, the capacitance and phase angle at 120 Hz are commonly used as figure of merits, in addition to the f_0, to evaluate the HF-ECs' performances. Key issues to design the electrode structure with high frequency response have already been explicitly discussed in previous review papers [1,34]. Since the maximum frequency response is linked with the RC time constant of the device, for a reasonably large capacitance C, its parasitic resistance (ESR) must be minimized. This requires that the electrode not only possesses high electronic conductivity and a large surface area, but also has a straightforward architecture which can facilitate the electrolyte ion to diffuse to the surface of the electrode rapidly with eliminated porous effect that results in the distributed nature of the charge storage. In addition, the contact resistance between the electrode and the current collector must be minimized as well.

Of the different nanocarbon structures, the edge-oriented VG network is positioned as being one of the superior candidates to meet the aforementioned requirements. In the VG structure (Figure 1e), multilayer graphene sheets grow vertically on a conductive substrate, forming an interconnected network, as shown by the scanning electron microscope (SEM) images in Figure 1b. The chemically strongly bonded VG sheet, the high intrinsic graphene conductivity, and the solid connection between VG sheets and the conducting substrate, all ensure a very low parasitic resistance. The fully exposed graphene edges and basal-planes can be easily impregnated by the electrolyte and the straight-forward channels created by the VG sheets endow rapid mass transport during electrochemical processes. Therefore, a much smaller ESR can be achieved as long as the conductivity of the electrolyte is reasonable. Moreover, as revealed by the transmission electron microscope (TEM) images (Figure 1c,d), the vertical sheets have a tapered shape with a thick base but a thin tip, along their wall edges there are distributed graphene steps or edges, which can offer abundant absorption sites for electrolyte ions. Thus, the exposed graphene edges and the easily accessed basal plane surfaces are expected to provide a promising capacitance density. All these merits collectively offer the possibility of creating VG-based HF-ECs with a minimized ESR and high level of charge storage capability. Therefore, it is understandable that the first kHz HF-EC was reported using VG for its electrodes. In this pioneering work, VG films were deposited on Ni foil by a plasma-enhanced vapor deposition (PECVD) technique and directly used as electrodes in a symmetric capacitor, achieving very fast frequency response with a f_0 of 15 kHz [2]. This unprecedented result demonstrated the feasibility of ECs working efficiently at the kHz range. Since then, steady progress has been achieved in this niche field by using a variety of electrode materials with favorable structures. In this mini-review, we summarize our contributions in the study of kHz HF-ECs by using VG grown on 3D scaffolds as electrodes. In Section 5, we also highlight a study of using cross-linked carbon nanofibers (CCNFs), derived from bacterial cellulose via rapid plasma pyrolysis, for HF-ECs. The highly conductive CCNFs electrode, in the absence of VG modification, can also offer a large surface area, rendering a capability of high levels of charge storage for HF-ECs.

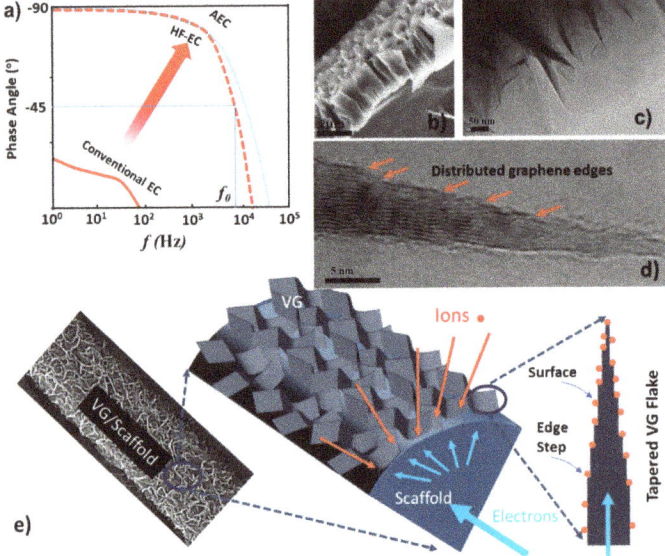

Figure 1. (**a**) Schematic showing the phase angle difference between conventional EC and AEC to appreciate their different frequency responses. HF-EC has a frequency response close to that of AEC, but with a much larger capacitance density. (**b**) SEM cross-sectional view of VG structure. (**c**) TEM image of vertical graphene sheets. (**d**) High-resolution TEM image of an individual sheet showing its tapered geometry with fully exposed graphene edges. (**e**) Schematic showing the VG network grown on a scaffold that offers high electronic conductivity and relatively large surface area provided by exposed graphene edges. Reproduced with permission from Ref. [1] for (**a**), Copyright 2017, Ref. [5] for (**b**), Copyright 2014, Ref. [6] for (**c**,**d**), Copyright 2016, and Ref. [9] for (**e**), Copyright 2019, Elsevier.

2. VG Structure Growth

VG electrodes are commonly grown on a conductive substrate in PECVD systems, in which hydrocarbons (e.g., CH_4 and C_2H_2) or fluorocarbons are used as carbon sources and reductive gases such as H_2 or NH_3 are used as etching reagents [35–37]. With plasma activation, the carbon-containing fragments are decomposed and deposited on the substrate, leading to the interconnected graphene sheets growing perpendicular to the substrate surface. There are different opinions on the VG growth mechanisms, including growth rate differences in vertical and lateral directions [38], crowding effects similar to aligned carbon nanotube growth [39], the effect of a vertical electric field across the plasma sheath [40], and the sp^2 bond bending upwards due to impinging planar island growth [41]. These factors may vary in importance in determining the vertical morphology, depending on growth conditions, especially the selection of substrates.

VG growth generally follows three major steps [36], as suggested by the SEM images (Figure 2) for VG deposition process on a Ni surface [42]: (a) During the first minute or so, after plasma deposition is initiated, a continuous laterally oriented multilayer-graphene or thin-graphite film composed of electrically well interconnected domains, is formed conformally along the nickel surface. In this buffer layer, graphene nuclei with their basal planes perpendicular to the substrate are incubated. (b,c) As the deposition proceeds, VG nucleation begins, and the growth of VG sheets starts along their basal plane edges, resulting in the gradually increased density of VG sheets. (d) The growth of a particular sheet terminates as its open edges close, but other active VG sheets continue to grow until a well-connected VG network forms in 10–15 min.

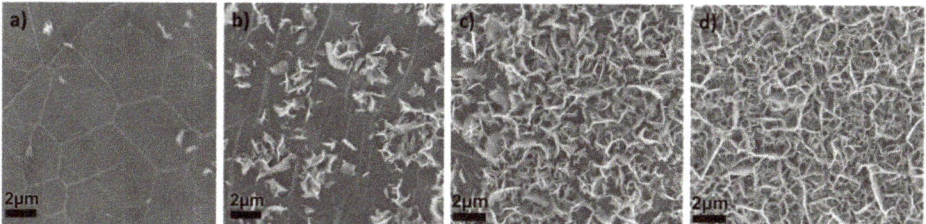

Figure 2. SEM images showing the surface evolution during VG deposition that suggest the three major steps of VG growth process: (**a**) the 1st minute, (**b**,**c**) in the following several minutes, and (**d**) after 15 min. Reproduced with permission from Ref. [42], Copyright 2016, Elsevier.

The height of the VG is generally limited within 1–2 μm, above which the coalescence of the individual sheet at the bottom will not be able to further increase the available surface area [43]. This restricted VG film height and the sheet density limit the achievable capacitance in terms of per unit area of the electrode footprint. In Ref. [2], the areal capacitance of VG on Ni foil at 120 Hz was only about 0.175 mF cm^{-2}. To increase the available graphene surface area and edge density for a given electrode footprint, we used conductive 3D scaffolds as the substrates for VG growth. As long as the 3D scaffold can be immerged in the plasma and the plasma sheath can be formed around the branches of the scaffold, the induced perpendicular field will trigger the reactive carbon fragments impinging onto the surface along the perpendicular direction, leading to the VG growth encircling around each branch in the scaffold. The results are an increased graphene sheet density with high surface area and an enhanced areal capacitance on a given electrode footprint.

The 3D scaffolds we used in a microwave based PECVD system include nickel foam [5,44], carbonized cellulose fiber paper [6,9] and carbon nanofiber film [22]. For a typical synthesis process, the scaffold loaded on a molybdenum holder was transferred into the growth chamber which was pumped down to 2×10^{-3} torr. As the sample holder reached the temperature of 750 °C, the scaffold was first cleaned by H$_2$ plasma, and then VG was grown under a certain of conditions for 10–15 min. It was estimated that the local temperature within the scaffold was more than ~1200 °C due to the plasma heating effects. Interestingly, when a cellulose fiber paper, such as tissue sheet or filter paper, was used as the scaffold, the carbonization of the substrate and VG growth were conducted simultaneously in the plasma atmosphere. This rapid plasma carbonization procedure, in contrast to the time-consuming pyrolysis process in the furnace, turned out to be the key step for developing a high-performance HF-EC electrode, which will be detailed in Section 4.

Aligned VG structures were also reported to be fabricated by wet chemistry-based methods [45,46], however, compared to the VG produced by the PECVD process, this kind of VG possesses a distinctive morphology and packing density. In particular, the lack of intimate contacts between the individual graphene sheet and between the substrate and the VG cause their relatively low conductivities, making them unsuitable as electrodes for HF-ECs.

3. VG on Ni Foam

Figure 3 is the schematic of synthesizing VG/Ni and freestanding VG foam electrodes [22]. SEM images at each step are presented here to show the morphology evolution. Using a bare Ni foam as the scaffold in a PECVD process (Figure 3a), carbon atoms first dissolve into nickel and then nucleate to form a thin graphite layer along the strut surface of the Ni foam, or G/Ni foam (Figure 3b). This is the buffer layer on which graphene nuclei with their basal planes perpendicular to the substrate are incubated. As deposition proceeds, the VG sheets start to grow along their basal plane edges and gradually transform into interconnected VG networks (Figure 3c). The underlying nickel foam can be chemically etched off in an acid solution to produce freestanding VG foam with greatly reduced weight (Figure 3d).

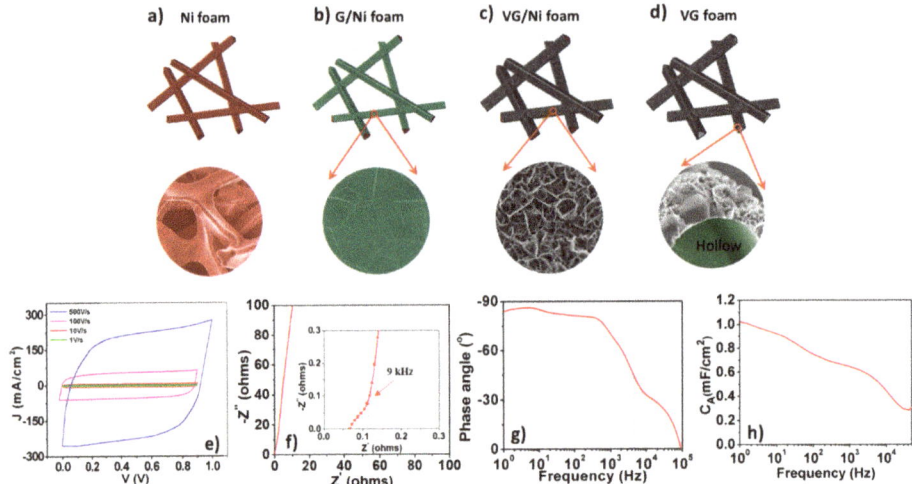

Figure 3. Schematic and the corresponding SEM images showing the morphology changes of (**a**) Ni foam, (**b**) G/Ni foam, (**c**) VG/Ni foam, and (**d**) freestanding VG foam after Ni is etched off. VG/Ni foam-based HF-EC performances: (**e**) CV profiles up to 500 V s^{-1}, (**f**) complex impedance spectrum (inset is the enlarged part at high frequency range. The electrode area is 1.7 cm^2.), (**g**) calculated electrode capacitance vs. frequency based on the RC model, and (**h**) the phase angle vs. frequency plot. Reproduced with permission from Ref. [42] for (**a–d**), Copyright 2016, and Ref. [5] for (**e–h**), Copyright 2014, Elsevier.

Here, VG/Ni foams were tested as electrodes for HF-EC. Without adding any binder and conductive additive, these electrodes were first wetted in 6 M KOH aqueous solution, and then assembled into symmetric ECs by using a 25 μm thick separator in a coin cell configuration (CR2016). As shown in Figure 3e, cyclic voltammetry (CV) curves collected within the potential range of 0–0.9 V, exhibit a quasi-rectangle shape with scan rates up to 500 V s^{-1}. The low resistivity and wide-open channels in VG/Ni foam could facilitate the migration of electrons and electrolyte ions, respectively, so an electrical double layer can be rapidly formed. As a result, the value of the specific capacitance for a single electrode can attain ~1.32 mF cm^{-2} at the scan rate of 1 V s^{-1} and remain at ~0.83 mF cm^{-2} at 500 V s^{-1}.

Electrochemical impedance spectroscopy (EIS) measurement was carried out over the frequency range from 100 kHz to 0.1 Hz. The Nyquist complex-plane impedance spectrum presented in Figure 3f has a nearly vertical line at low frequency. Above a knee frequency of ~9 kHz, the line becomes more tilted and an intersection with the real axis at around 45° suggests a typical porous electrode behavior, caused by the hierarchical structure of VG/Ni foam. The spectrum at frequencies lower than the knee frequency can be commonly modeled as an ideal capacitor with an ESR. Using the formula $C = -1/(2\pi f Z'')$, where f is the frequency and Z'' is the imaginary part of the impedance, the areal capacitance of the electrode can be calculated, and the result is plotted in Figure 3g. The cell gives specific capacitances of ~0.72 and ~0.64 mF cm^{-2} at 120 Hz and 1 kHz, respectively. The Bode diagram in Figure 3h shows that the phase angle of the cell at 120 Hz is about −82° and the f_0 is around 4 kHz. With a cutoff in kHz scope, the VG/Ni foam electrode has a much higher areal capacitance than the electrodes made from VG on flat foils [2,4].

4. VG on Carbon Fiber Sheets

Although a reasonable areal capacitance could be obtained using VG/Ni foam electrode, its volumetric density is considerably lower due to the relatively low area:volume ratio of Ni foam. To address this problem, we further employed carbon nanofiber (CNF) films as the substrates for

VG growth [22]. Commercially available CNFs were first dispersed in ethanol solution with the assistance of surfactants and then spin-coated onto Ni foil to form a thin layer with macroporous structure after drying. Subsequently, VG was successfully grown into this porous layer with graphene nanosheets wrapped around each CNF in a PECVD system, forming a highly conductive electrode with a large surface area. In an aqueous electrolyte cell, f_0 was found to be as high as 22 kHz and the areal capacitance for a single electrode was about 0.37 mF cm^{-2} at 120 Hz.

More interesting works focus on VG growth on carbonized cellulose microfibers (CMFs) and their HF-EC applications [6,9]. Different from the reported cellulose derived carbon materials for electrochemical applications [47–50], the freestanding VG/CMF electrodes used in our studies were synthesized in a one-step PECVD process, where cellulose fiber sheets were rapidly pyrolyzed by the high-temperature plasma and VG was grown on the developing CMFs simultaneously. For instance, by using cellulose wiper (Kimwipes^{TM}) as the raw material, a thin (~10 μm) and freestanding electrode was obtained after the PECVD process, which gave satisfying performances in terms of both frequency response and specific capacitance. Furthermore, these thin layers with hierarchical macro-pore structure can be further stacked to obtain thick electrodes with much higher areal capacitance without compromising their frequency response and volumetric capacitance. Apparently, compared to the VG/metal foam structure, this kind of electrode architecture is more promising in practical use.

The wood fiber derived unwoven cellulose paper consists of abundant ribbon-shaped cellulose fiber bundles, and these bundles have a relatively compact structure in the microscale. During the high-temperature PECVD process, the majority of O and H elements are removed, and cellulose fiber is converted into amorphous carbon fiber and further partially graphitized. A large hollow space between the neighboring carbon fiber bundles can be preserved. As demonstrated in Figure 4a, pyrolysis in PECVD causes cellulose texture to have an anisotropic contraction and the sheet size is reduced by a factor of 5–6 after being converted into VG/CMF. Composition analysis shows that only a trace amount of oxygen element, mainly in the C–O bonding form, remains in CMF. The X-ray diffraction (XRD) patterns of cellulose fiber and VG/CMF in Figure 4b show the differences in their crystalline structures. The pristine cellulose fiber, consisting of cellulose I$_β$, has a monoclinic unit cell. The observed five peaks in the XRD pattern can be well indexed to cellulose I$_β$ (101), (10$\bar{1}$), (021), (002), and (040) planes [51]. In contrast, the VG/CMF only has one diffraction peak at ~26.2°, corresponding to the characteristic (002) plane of graphite. SEM images (Figure 4c,d) reveal a maze-like network composed of VG sheets wrapped around CMFs, with plenty of exposed graphene edges and straightforward channels. By using the thin and freestanding VG/CMF sheets as electrodes and KOH aqueous solution as electrolyte, the resulting HF-ECs exhibited superior performances with f_0 above the kHz, areal capacitance of 0.6–1.5 mF cm^{-2} (depending on the electrode thickness) and volumetric capacitance of 0.6 mF cm^{-3} in terms of a single electrode.

Even though aqueous based kHz HF-ECs have been widely reported in the literature, their low voltage ratings, 0.8–0.9 V restricted by water decomposition, limit their practical applications. Organic electrolytes have larger voltage ratings (2.5–3.5 V), rendering them a better electrolyte option for practical HF-ECs in spite of their relatively low conductivity. Herein, VG/CMF electrode based organic cells were also studied by using 1 M tetra-ethylammonium tetrafluoroborate (TEABF$_4$) in anhydrous acetonitrile (AN) solution as the electrolyte. The CV curve (Figure 4e) indicates that the potential window of the organic HF-EC is significantly widened by a factor of ~3 in comparison with that of the inorganic one and maintains the desired quasi-rectangular shape at the scan rate of 1000 V s^{-1}, reflecting its high-speed capability. In addition, from the Nyquist plot of this organic cell (Figure 4f), it is observed that a small semicircle and a near 45° linear region appear in the high frequency scope and a near vertical part shows up below the knee frequency of 3.8 kHz, indicating that this impedance plot can be simulated by a series RC model as described above. By using equations based on this model, the areal capacitance of single electrode was calculated (Figure 4g) and found to be ~0.49 mF cm^{-2} at 120 Hz. From its Bode diagram (Figure 4g), the organic cell exhibits a phase angle of −80.4° at 120 Hz and an f_0 of 1.3 kHz. The impedance spectrum was further analyzed by introducing complex capacitance $C = C' - jC''$ (where C' is the accessible capacitance at the corresponding frequency and C'' corresponds to energy dissipation) and the results are presented in Figure 4h. The modeled capacitance (C') agrees with the result shown in Figure 4g. The value of C'' reaches a maximum at a frequency of 1.3 kHz, defining a relaxation time constant (τ_0) of 0.8 ms, in accordance with the characteristic frequency.

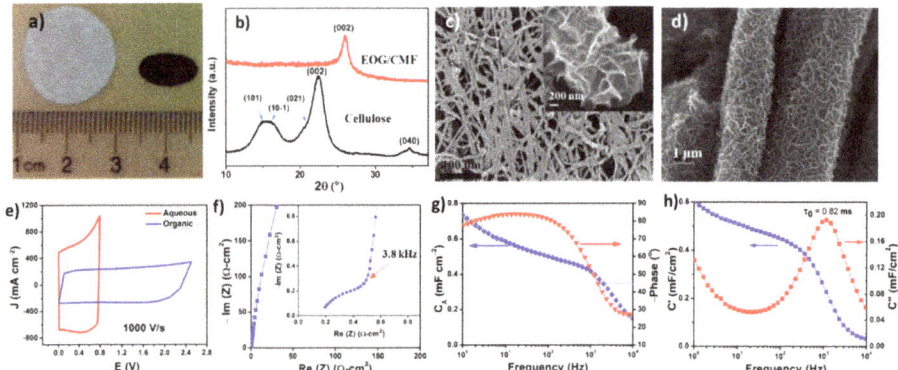

Figure 4. (**a**) Photo showing the cotton round (cellulose sheet, the white one) and the VG/CMF sheet (the black one). (**b**) XRD patterns of the cellulose and the VG/CMF sheets. (**c**,**d**) SEM images of VG/CMF at different magnifications. Electrochemical performances of VG/CMF electrode: (**e**) CV curves at 1000 V s^{-1} rate in aqueous and organic electrolyte cells. (**f**) Nyquist impedance spectrum with the inset showing the zoom-in high frequency range, (**g**) *RC* model-derived electrode's areal capacitance vs. frequency plot and its Bode diagram, and (**h**) real and imaginary components of the complex electrode capacitances in an organic cell. Reproduced with permission from Ref. [6] for (**a**–**c**), Copyright 2016, and Ref. [9] for (**e**–**h**), Copyright 2019, Elsevier.

As a proof of concept, the assembled VG/CMF-based HF-EC with organic electrolyte was further tested for ripple current filtering in the 60 Hz line-frequency AC/DC conversion. The circuit diagram is described in Figure 5a, a 60 Hz sinusoidal wave voltage with a 4 V amplitude (Figure 5b) is simulated as the input which passes through a full-bridge rectifier and an HF-EC filter capacitor in turn before being supplied to a load. As shown in Figure Figure 5c, compared to the output voltage profile collected without using the filter capacitor, a near ripple-free 2.8 V DC voltage applied to the load is obtained. This clearly demonstrates that, with their larger capacitance densities, HF-ECs are promising to replace bulky AECs for AC line-filtering in electronics and power applications in the future.

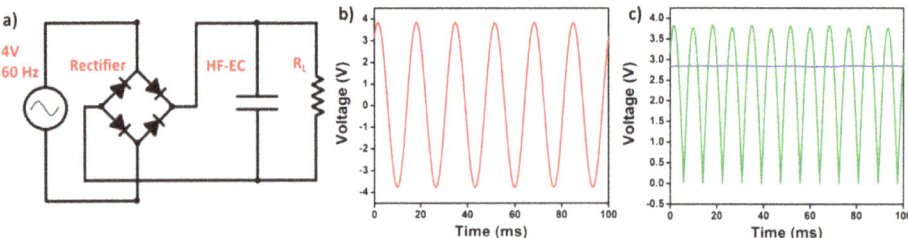

Figure 5. Demonstration of AC line filtering function of HF-EC using VG/CMF electrodes: (**a**) AC/DC conversion circuit diagram, (**b**) a 60 Hz sine wave signal with 4 V amplitude used as the input, and (**c**) the output from the rectifier with (blue line) and without (green curve) filtering capacitor. Reproduced with permission from Ref. [9], Copyright 2019, Elsevier.

5. Cross-Linked Carbon Nanofiber Derived from Bacterial Cellulose Aerogel

In addition to VG based electrode materials, we also studied cross-linked carbon nanofibers (CCNFs) to simplify the electrode structure but maintain their excellent performances. The CCNFs were obtained by rapid pyrolysis of bacterial cellulose (BC) precursor (Figure 6a) in the PECVD system. BC pellicles are comprised of cellulose microfibrils secreted bacterially. These cellulose microfibrils first are bound into bundles and then are woven into nanoribbons which further branch into a three dimensional web [52,53]. In our research, BC pellicles were synthesized using Kombucha strains by a fermentation process [54]. After chemical cleaning to eliminate bacterial cells in the BC pellicles and freeze-drying processes, BC aerogel was produced, which contained less than 1 wt % of its hydrogel counterpart.

As a precursor for carbon nanofibers, BC is preferred over other natural celluloses due to its smaller fiber size (around 10–50 nm, Figure 6b,c), higher degree of crystallinity and purity (Figure 6d,e) [55–58]. Exclusively, unlike cellulose fibers derived from plants that are not interconnected, BC with branched structure (Figure 6c) can yield cross-linked CNFs (Figure 6f) after pyrolysis. These downsized carbon nanofibers with cross-linked structure not only decrease the inner contact resistance, but also provide a larger surface area compared to other microscale fibers, and are expected to have a larger capacitance density.

As we know, the pyrolysis processes for polymer and cellulose precursors are commonly conducted in an inert ambient atmosphere at high temperatures for several hours [59,60]. Micropore activation, through physical decomposition and chemical etching, is a common practice for achieving a large surface area [58]. Figure 6g compares the complex impedance spectra of CCNFs produced by rapid plasma pyrolysis in the PECVD system and conventional thermal pyrolysis in a tube furnace [61]. Although the traditional thermal pyrolysis method can dramatically enlarge the specific surface area of CCNFs by introducing a large amount of micropores, the frequency response of CCNFs is significantly restrained, making them unsuitable for HF-ECs. In contrast, CCNF film fabricated by the rapid plasma pyrolysis process can avoid this microporous structure, which is a key factor for its success in HF-EC application. The nitrogen absorption–desorption result of this CCNF gives a specific surface area of 57.5 $m^2\ g^{-1}$ and a pore volume of 0.374 $cm^3\ g^{-1}$ with a minimum pore diameter of 3.8 nm. The meso- and macro-pores allow a rapid transportation of electrolyte ions throughout the electrode mesh with a low diffusion resistance, and the cross-linked structure of CNFs guarantees a high electronic conductivity. Consequently, this web-like electrode engenders a fast frequency response with improved specific capacitances.

CCNF freestanding electrodes with different thickness (10, 20, and 60 µm) were first studied in KOH aqueous electrolyte by assembling them into coin cells. For example, the Bode diagram of the 20 µm electrode (Figure 6h) shows that the absolute value of the phase angle stays above 80° with frequency increasing to a few hundred Hz. In particular, the 120 Hz phase angle is around −82° and the f_0 is close to 3.3 kHz. The value of the areal capacitance for a single electrode at 120 Hz is proportional to the thickness of the electrode, while f_0 varies inversely ((1.51, 4,1), (2.98, 3.3) and (4.50 mF cm^{-2}, 1.3 kHz) for 10, 20, and 60 µm electrode, respectively). For the two thinner electrodes, a volumetric capacitance of ~1.50 mF cm^{-3} is achieved. Figure 6i shows the results of cycling stability for the 10 µm electrodes. After 100,000 continuous cycles with full charge and discharge, about 95% of its initial capacity is maintained. These CCNF electrodes exhibited extraordinary performance in terms of both areal capacitance and frequency response.

To broaden its applicable potential window, 10 µm CCNF electrodes were further studied using 1 M $TEABF_4$/AN electrolyte, in which the cell can work at an elevated voltage up to 3.5 V (Figure 7a). Other promising electrochemical results such as a knee frequency of 3.0 kHz, a phase angle of −80°, and an areal capacitance of 0.51 mF cm^{-2} for a single electrode at 120 Hz (corresponding to volumetric capacitance of 0.51 F cm^{-3}) and the characteristic frequency of 1.8 kHz well prove that these CCNFs are one of a number of excellent candidates for high voltage AC line filtering ECs.

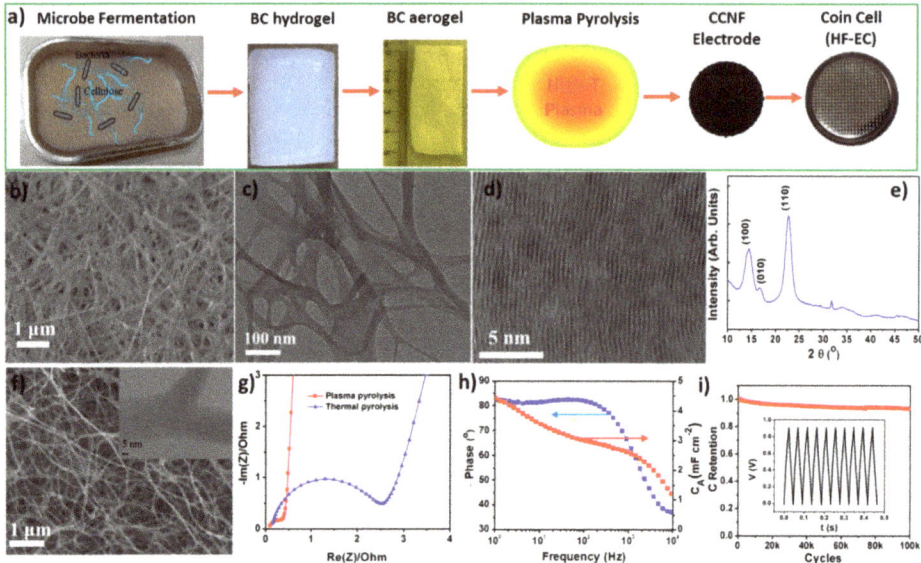

Figure 6. (**a**) Schematic showing the BC-derived CCNF fabrication processes and their electrochemical test. (**b**) SEM image of BC aerogel. (**c**) TEM image showing the cross-linked structure of BC nanofibers. (**d**) High-resolution TEM image and (**e**) XRD pattern showing the high crystalline of BC. (**f**) SEM image of CCNF showing the macroporous structure. The inset TEM image shows branched CNF. (**g**) Comparison of complex impedance spectra between CCNF produced by rapid plasma pyrolysis and conventional thermal pyrolysis. (**h**) Plots of phase angle and electrode capacitance vs. frequency for a 20 μm thick CCNF electrode. (**i**) Galvanostatic charge-discharge stability result at a current density of 50 mA cm^{-2} for a 10 μm CCNF electrode, with the inset as a section of the C-D curve. Reproduced with permission from Ref. [13] Copyright 2017, Elsevier.

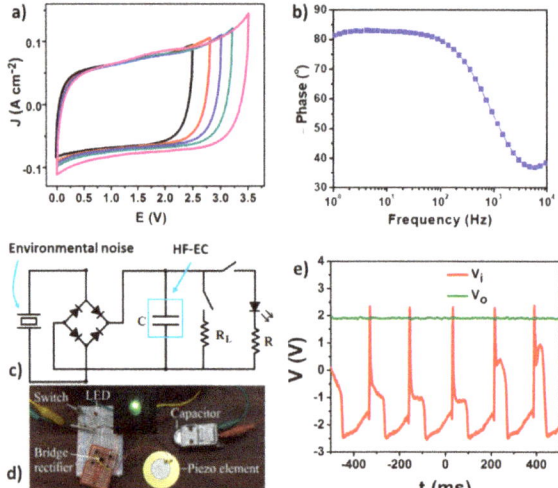

Figure 7. (**a**) CV profiles within different potential windows at scan rate of 100 Vs^{-1} and (**b**) Bode plot for the organic cell. (**c**) The diagram and (**d**) photo of the circuit used for pulse energy scavenge and storage. (**e**) The scavenged pulse V_i (red curve) and the DC output V_o (green curve) after being filtered by the HF-EC to a micro-power load. Reproduced with permission from Ref. [13] Copyright 2017, Elsevier.

Besides filtering applications, HF-ECs, if their self-discharge and leakage current can be minimized, may also be suitable for pulse energy storage in an energy scavenge system. There are strong interests in using piezoelectric or triboelectric mechanisms to harvest the environmental noise and vibration and convert them into electricity for self-powered autonomous sensors. Since these mechanical energy sources typically vibrate at tens or hundreds of Hz [62–64], HF-ECs, as compact energy storage devices, are needed to efficiently store these harvested pulse powers. As a demonstration in our study, a piezoelectric element was used to generate pulses from vibrations. The testing circuit diagram and photo of these elements are respectively shown in Figure 7c,d. Pulses picked up by a piezoelectric microgenerator are first rectified by a full-bridge and then stored in the HF-EC. This stored energy is further used to power a micro-sensor and a green light-emitting diode (LED) is used to simulate a high-power pulse load. In Figure 7e, the voltage pulses (V_i) from finger-tapping of the piezoelectric element and the DC output (V_o) to the micro-load are presented. In addition to providing a constant current to the micro-power load, this HF-EC can even drive a high-power pulse load by turning on a green LED for a short period. These preliminary results prove that our fabricated CCNF based HF-ECs are promising for ripple current filtering and pulse energy harvest and storage applications.

6. Conclusions

By designing and synthesizing different types of carbon-based electrodes, great progress is being steadily made toward practical HF-EC applications, with a goal of very compact EC devices acting as both energy storages and filtering capacitors. In this mini-review, based on our selected works in this field, the characteristics and performances of several representative VG based materials are discussed, with their pros and cons clearly described. The substrates used for growing VG have been proved to be critical in determining the final specific capacitance and the speed of frequency response of HF-ECs. Compared to frequently used metal foams, carbon fibers with a three-dimensional configuration are more suitable for HF-ECs in terms of their light weight, reduced volume, high specific surface area, and resistance to corrosive electrolytes. Benefiting from the high efficiency of the PECVD pyrolysis process, more structural engineering works for the substrates could be considered in the future. For instance, the metal foams normally have a very large pore size (tens to hundreds of micrometer) which not only reduces the available sites for VG growth, but also decreases the volumetric capacitance of the electrochemical devices. Filling part of the pores with other materials such as polymers or carbon fibers before VG growth may alleviate these problems. In Section 5, CCNFs are shown to be superior in electrochemical performances even to VG/CFs, ascribed to their cross-linked structure and highly available surface area. This unique architecture constructed by the branched nanofibers could provide fast ion and electron transportation pathways, resulting in a more desirable frequency response as well as specific capacitance and cycling stability in both inorganic and organic electrolytes. Furthermore, the successful demonstrations in AC line filtering and pulse energy harvesting and storage give them a promising prospect in practical applications. These results also provide a reference for optimizing the performance of carbon-based electrodes by introducing effective cross-linked structures.

Last but not the least, although some HF-ECs made from different electrodes have been proved to possess excellent electrochemical performances, their applicable potential windows are still restricted by the electrolytes' decomposition voltages (~0.8–0.9 V for water-based electrolyte and ~3.0–3.5 V for organic based electrolytes), which inevitably limit their application areas. The rated voltage can be elevated by the stacking process, but the capacitance as well as the frequency response will be decreased dramatically, making them uncompetitive in practical use. How to increase their rated voltages without compromising their frequency response speed, becomes a tough problem needing to be addressed urgently. Stacking design, packaging configurations, and minimizing the contact resistances between single cells are very important in solving this issue. With more efforts being currently carried out in this research area, we believe that significant achievements will soon be implemented.

Funding: This work is supported by National Science Foundation (1611060, 1820098).

Conflicts of Interest: The authors declare no conflict of interest.

References

1. Fan, Z.; Islam, N.; Bayne, S.B. Towards kilohertz electrochemical capacitors for filtering and pulse energy harvesting. *Nano Energy* **2017**, *39*, 306–320. [CrossRef]
2. Miller, J.R.; Outlaw, R.A.; Holloway, B.C. Graphene double-layer capacitor with ac line-filtering performance. *Science* **2010**, *329*, 1637–1639. [CrossRef] [PubMed]
3. Sheng, K.; Sun, Y.; Li, C.; Yuan, W.; Shi, G. Ultrahigh-rate supercapacitors based on eletrochemically reduced graphene oxide for ac line-filtering. *Sci. Rep.* **2012**, *2*, 247. [CrossRef] [PubMed]
4. Cai, M.; Outlaw, R.A.; Butler, S.M.; Miller, J.R. A high density of vertically-oriented graphenes for use in electric double layer capacitors. *Carbon* **2012**, *50*, 5481–5488. [CrossRef]
5. Ren, G.; Pan, X.; Bayne, S.; Fan, Z. Kilohertz ultrafast electrochemical supercapacitors based on perpendicularly-oriented graphene grown inside of nickel foam. *Carbon* **2014**, *71*, 94–101. [CrossRef]
6. Ren, G.; Li, S.; Fan, Z.-X.; Hoque, M.N.F.; Fan, Z. Ultrahigh-rate supercapacitors with large capacitance based on edge oriented graphene coated carbonized cellulous paper as flexible freestanding electrodes. *J. Power Sources* **2016**, *325*, 152–160. [CrossRef]
7. Miller, J.R.; Outlaw, R.A. Vertically-oriented graphene electric double layer capacitor designs. *J. Electrochem. Soc.* **2015**, *162*, A5077–A5082. [CrossRef]
8. Zhang, M.; Zhou, Q.; Chen, J.; Yu, X.; Huang, L.; Li, Y.; Li, C.; Shi, G. An ultrahigh-rate electrochemical capacitor based on solution-processed highly conductive PEDOT: PSS films for AC line-filtering. *Energy Environ. Sci.* **2016**, *9*, 2005–2010. [CrossRef]
9. Islam, N.; Hoque, M.N.F.; Li, W.; Wang, S.; Warzywoda, J.; Fan, Z. Vertically edge-oriented graphene on plasma pyrolyzed cellulose fibers and demonstration of kilohertz high-frequency filtering electrical double layer capacitors. *Carbon* **2019**, *141*, 523–530. [CrossRef]
10. Kossyrev, P. Carbon black supercapacitors employing thin electrodes. *J. Power Sources* **2012**, *201*, 347–352. [CrossRef]
11. Strauss, V.; Anderson, M.; Turner, C.L.; Kaner, R.B. Fast response electrochemical capacitor electrodes created by laser-reduction of carbon nanodots. *Mater. Today Energy* **2019**, *11*, 114–119. [CrossRef]
12. Zhao, M.; Nie, J.; Li, H.; Xia, M.; Liu, M.; Zhang, Z.; Liang, X.; Qi, R.; Wang, Z.L.; Lu, X. High-frequency supercapacitors based on carbonized melamine foam as energy storage devices for triboelectric nanogenerators. *Nano Energy* **2019**, *55*, 447–453. [CrossRef]
13. Islam, N.; Li, S.; Ren, G.; Zu, Y.; Warzywoda, J.; Wang, S.; Fan, Z. High-frequency electrochemical capacitors based on plasma pyrolyzed bacterial cellulose aerogel for current ripple filtering and pulse energy storage. *Nano Energy* **2017**, *40*, 107–114. [CrossRef]
14. Yoo, Y.; Kim, S.; Kim, B.; Kim, W. 2.5 V compact supercapacitors based on ultrathin carbon nanotube films for AC line filtering. *J. Mater. Chem. A* **2015**, *3*, 11801–11806. [CrossRef]
15. Rangom, Y.; Tang, X.S.; Nazar, L.F. Carbon Nanotube-Based Supercapacitors with Excellent ac Line Filtering and Rate Capability via Improved Interfacial Impedance. *ACS Nano* **2015**, *9*, 7248–7255. [CrossRef] [PubMed]
16. Li, Q.; Sun, S.; Smith, A.D.; Lundgren, P.; Fu, Y.; Su, P.; Xu, T.; Ye, L.; Sun, L.; Liu, J. Compact and low loss electrochemical capacitors using a graphite/carbon nanotube hybrid material for miniaturized systems. *J. Power Sources* **2019**, *412*, 374–383. [CrossRef]
17. Lin, J.; Zhang, C.; Yan, Z.; Zhu, Y.; Peng, Z.; Hauge, R.H.; Natelson, D.; Tour, J.M. 3-Dimensional graphene carbon nanotube carpet-based microsupercapacitors with high electrochemical performance. *Nano Lett.* **2013**, *13*, 72–78. [CrossRef]
18. Yoo, Y.; Kim, M.S.; Kim, J.K.; Kim, Y.S.; Kim, W. Fast-response supercapacitors with graphitic ordered mesoporous carbons and carbon nanotubes for AC line filtering. *J. Mater. Chem. A* **2016**, *4*, 5062–5068. [CrossRef]
19. Eftekhari, A.; Fan, Z. Ordered mesoporous carbon and its applications for electrochemical energy storage and conversion. *Mater. Chem. Front.* **2017**, *1*, 1001–1027. [CrossRef]

20. Zhang, Z.; Liu, M.; Tian, X.; Xu, P.; Fu, C.; Wang, S.; Liu, Y. Scalable fabrication of ultrathin free-standing graphene nanomesh films for flexible ultrafast electrochemical capacitors with AC line-filtering performance. *Nano Energy* **2018**, *50*, 182–191. [CrossRef]
21. Cai, M.; Outlaw, R.A.; Quinlan, R.A.; Premathilake, D.; Butler, S.M.; Miller, J.R. Fast Response, vertically oriented graphene nanosheet electric double layer capacitors synthesized from C(2)H(2). *ACS Nano* **2014**, *8*, 5873–5882. [CrossRef] [PubMed]
22. Islam, N.; Warzywoda, J.; Fan, Z. Edge-Oriented Graphene on Carbon Nanofiber for High-Frequency Supercapacitors. *Nanomicro Lett.* **2018**, *10*, 9. [CrossRef] [PubMed]
23. Premathilake, D.; Outlaw, R.A.; Parler, S.G.; Butler, S.M.; Miller, J.R. Electric double layer capacitors for ac filtering made from vertically oriented graphene nanosheets on aluminum. *Carbon* **2017**, *111*, 231–237. [CrossRef]
24. Premathilake, D.; Outlaw, R.A.; Quinlan, R.A.; Parler, S.G.; Butler, S.M.; Miller, J.R. Fast Response, Carbon-Black-Coated, Vertically-Oriented Graphene Electric Double Layer Capacitors. *J. Electrochem. Soc.* **2018**, *165*, A924–A931. [CrossRef]
25. Kang, Y.J.; Yoo, Y.; Kim, W. 3-V Solid-State Flexible Supercapacitors with Ionic-Liquid-Based Polymer Gel Electrolyte for AC Line Filtering. *ACS Appl. Mater. Interfaces* **2016**, *8*, 13909–13917. [CrossRef] [PubMed]
26. Gao, H.; Li, J.; Miller, J.R.; Outlaw, R.A.; Butler, S.; Lian, K. Solid-state electric double layer capacitors for ac line-filtering. *Energy Storage Mater.* **2016**, *4*, 66–70. [CrossRef]
27. Laszczyk, K.U.; Kobashi, K.; Sakurai, S.; Sekiguchi, A.; Futaba, D.N.; Yamada, T.; Hata, K. Lithographically Integrated Microsupercapacitors for Compact, High Performance, and Designable Energy Circuits. *Adv. Energy Mater.* **2015**, *5*, 1500741. [CrossRef]
28. Pan, X.; Ren, G.; Hoque, M.N.F.; Bayne, S.; Zhu, K.; Fan, Z. Fast Supercapacitors Based on Graphene-Bridged V2O3/VOx Core–Shell Nanostructure Electrodes with a Power Density of 1 MW kg^{-1}. *Adv. Mater. Interfaces* **2014**, *1*, 1400398. [CrossRef]
29. Yang, P.; Chao, D.; Zhu, C.; Xia, X.; Zhang, Y.; Wang, X.; Sun, P.; Tay, B.K.; Shen, Z.X.; Mai, W. Ultrafast-Charging Supercapacitors Based on Corn-Like Titanium Nitride Nanostructures. *Adv. Sci.* **2015**, *3*, 1500299. [CrossRef]
30. Liu, W.; Lu, C.; Wang, X.; Tay, R.Y.; Tay, B.K. High-performance microsupercapacitors based on two-dimensional graphene/manganese dioxide/silver nanowire ternary hybrid film. *ACS Nano* **2015**, *9*, 1528–1542. [CrossRef]
31. Kurra, N.; Jiang, Q.; Syed, A.; Xia, C.; Alshareef, H.N. Micro-Pseudocapacitors with Electroactive Polymer Electrodes: Toward AC-Line Filtering Applications. *ACS Appl. Mater. Interfaces* **2016**, *8*, 12748–12755. [CrossRef] [PubMed]
32. Gund, G.S.; Park, J.H.; Harpalsinh, R.; Kota, M.; Shin, J.H.; Kim, T.-i.; Gogotsi, Y.; Park, H.S. MXene/Polymer Hybrid Materials for Flexible AC-Filtering Electrochemical Capacitors. *Joule* **2018**, *3*, 164–176. [CrossRef]
33. Islam, N.; Wang, S.; Warzywoda, J.; Fan, Z. Fast supercapacitors based on vertically oriented MoS$_2$ nanosheets on plasma pyrolyzed cellulose filter paper. *J. Power Sources* **2018**, *400*, 277–283. [CrossRef]
34. Eftekhari, A. The mechanism of ultrafast supercapacitors. *J. Mater. Chem. A* **2018**, *6*, 2866–2876. [CrossRef]
35. Hiramatsu, M.; Hori, M. *Carbon Nanowalls: Synthesis and Emerging Applications*; Springer Science & Business Media: New York, NY, USA, 2010.
36. Bo, Z.; Mao, S.; Han, Z.J.; Cen, K.; Chen, J.; Ostrikov, K.K. Emerging energy and environmental applications of vertically-oriented graphenes. *Chem. Soc. Rev.* **2015**, *44*, 2108–2121. [CrossRef]
37. Zhang, Z.; Lee, C.S.; Zhang, W. Vertically aligned graphene nanosheet arrays: synthesis, properties and applications in electrochemical energy conversion and storage. *Adv. Energy Mater.* **2017**, *7*, 1700678. [CrossRef]
38. Hiramatsu, M.; Shiji, K.; Amano, H.; Hori, M. Fabrication of vertically aligned carbon nanowalls using capacitively coupled plasma-enhanced chemical vapor deposition assisted by hydrogen radical injection. *Appl. Phys. Lett.* **2004**, *84*, 4708–4710. [CrossRef]
39. Teii, K.; Shimada, S.; Nakashima, M.; Chuang, A.T. Synthesis and electrical characterization of n-type carbon nanowalls. *J. Appl. Phys.* **2009**, *106*, 084303. [CrossRef]
40. Zhu, M.; Wang, J.; Holloway, B.C.; Outlaw, R.; Zhao, X.; Hou, K.; Shutthanandan, V.; Manos, D.M. A mechanism for carbon nanosheet formation. *Carbon* **2007**, *45*, 2229–2234. [CrossRef]

41. Malesevic, A.; Vitchev, R.; Schouteden, K.; Volodin, A.; Zhang, L.; Van Tendeloo, G.; Vanhulsel, A.; Van Haesendonck, C. Synthesis of few-layer graphene via microwave plasma-enhanced chemical vapour deposition. *Nanotechnology* **2008**, *19*, 305604. [CrossRef]
42. Ren, G.; Hoque, M.N.F.; Liu, J.; Warzywoda, J.; Fan, Z. Perpendicular edge oriented graphene foam supporting orthogonal TiO_2 (B) nanosheets as freestanding electrode for lithium ion battery. *Nano Energy* **2016**, *21*, 162–171. [CrossRef]
43. Pan, X.; Zhu, K.; Ren, G.; Islam, N.; Warzywoda, J.; Fan, Z. Electrocatalytic properties of a vertically oriented graphene film and its application as a catalytic counter electrode for dye-sensitized solar cells. *J. Mater. Chem. A* **2014**, *2*, 12746–12753. [CrossRef]
44. Ren, G.; Hoque, M.N.F.; Pan, X.; Warzywoda, J.; Fan, Z. Vertically aligned VO_2 (B) nanobelt forest and its three-dimensional structure on oriented graphene for energy storage. *J. Mater. Chem. A* **2015**, *3*, 10787–10794. [CrossRef]
45. Yoon, Y.; Lee, K.; Kwon, S.; Seo, S.; Yoo, H.; Kim, S.; Shin, Y.; Park, Y.; Kim, D.; Choi, J.Y.; et al. Vertical alignments of graphene sheets spatially and densely piled for fast ion diffusion in compact supercapacitors. *ACS Nano* **2014**, *8*, 4580–4590. [CrossRef] [PubMed]
46. Zhu, J.; Sakaushi, K.; Clavel, G.; Shalom, M.; Antonietti, M.; Fellinger, T.-P. A general salt-templating method to fabricate vertically aligned graphitic carbon nanosheets and their metal carbide hybrids for superior lithium ion batteries and water splitting. *J. Am. Chem. Soc.* **2015**, *137*, 5480–5485. [CrossRef] [PubMed]
47. Jiang, L.; Nelson, G.W.; Kim, H.; Sim, I.N.; Han, S.O.; Foord, J.S. Cellulose-Derived Supercapacitors from the Carbonisation of Filter Paper. *ChemistryOpen* **2015**, *4*, 586–589. [CrossRef] [PubMed]
48. Zhang, L.; Zhu, P.; Zhou, F.; Zeng, W.; Su, H.; Li, G.; Gao, J.; Sun, R.; Wong, C.P. Flexible Asymmetrical Solid-State Supercapacitors Based on Laboratory Filter Paper. *ACS Nano* **2016**, *10*, 1273–1282. [CrossRef] [PubMed]
49. Li, S.; Ren, G.; Hoque, M.N.F.; Dong, Z.; Warzywoda, J.; Fan, Z. Carbonized cellulose paper as an effective interlayer in lithium-sulfur batteries. *Appl. Surf. Sci.* **2017**, *396*, 637–643. [CrossRef]
50. Li, S.; Fan, Z. Nitrogen-doped carbon mesh from pyrolysis of cotton in ammonia as binder-free electrodes of supercapacitors. *Microporous Mesoporous Mater.* **2019**, *274*, 313–317. [CrossRef]
51. Garvey, C.J.; Parker, I.H.; Simon, G.P. On the interpretation of X-ray diffraction powder patterns in terms of the nanostructure of cellulose I fibres. *Macromol. Chem. Phys.* **2005**, *206*, 1568–1575. [CrossRef]
52. Esa, F.; Tasirin, S.M.; Rahman, N.A. Overview of Bacterial Cellulose Production and Application. *Agric. Agric. Sci. Procedia* **2014**, *2*, 113–119. [CrossRef]
53. Hu, W.; Chen, S.; Yang, J.; Li, Z.; Wang, H. Functionalized bacterial cellulose derivatives and nanocomposites. *Carbohydr. Polym.* **2014**, *101*, 1043–1060. [CrossRef] [PubMed]
54. Zhu, C.; Li, F.; Zhou, X.; Lin, L.; Zhang, T. Kombucha-synthesized bacterial cellulose: Preparation, characterization, and biocompatibility evaluation. *J. Biomed. Mater. Res. Part A* **2014**, *102*, 1548–1557. [CrossRef] [PubMed]
55. Huang, H.; Tang, Y.; Xu, L.; Tang, S.; Du, Y. Direct formation of reduced graphene oxide and 3D lightweight nickel network composite foam by hydrohalic acids and its application for high-performance supercapacitors. *ACS Appl. Mater. Interfaces* **2014**, *6*, 10248–10257. [CrossRef] [PubMed]
56. Wu, Z.Y.; Li, C.; Liang, H.W.; Chen, J.F.; Yu, S.H. Ultralight, flexible, and fire-resistant carbon nanofiber aerogels from bacterial cellulose. *Angew. Chem.* **2013**, *125*, 2997–3001. [CrossRef]
57. Li, S.; Mou, T.; Ren, G.; Warzywoda, J.; Wei, Z.; Wang, B.; Fan, Z. Gel based sulfur cathodes with a high sulfur content and large mass loading for high-performance lithium–sulfur batteries. *J. Mater. Chem. A* **2017**, *5*, 1650–1657. [CrossRef]
58. Li, S.; Warzywoda, J.; Wang, S.; Ren, G.; Fan, Z. Bacterial cellulose derived carbon nanofiber aerogel with lithium polysulfide catholyte for lithium–sulfur batteries. *Carbon* **2017**, *124*, 212–218. [CrossRef]
59. Wang, X.; Kong, D.; Zhang, Y.; Wang, B.; Li, X.; Qiu, T.; Song, Q.; Ning, J.; Song, Y.; Zhi, L. All-biomaterial supercapacitor derived from bacterial cellulose. *Nanoscale* **2016**, *8*, 9146–9150. [CrossRef]
60. Chen, L.-F.; Huang, Z.-H.; Liang, H.-W.; Gao, H.-L.; Yu, S.-H. Three-Dimensional Heteroatom-Doped Carbon Nanofiber Networks Derived from Bacterial Cellulose for Supercapacitors. *Adv. Funct. Mater.* **2014**, *24*, 5104–5111. [CrossRef]
61. Islam, N.; Hoque, M.N.F.; Zu, Y.; Wang, S.; Fan, Z. Carbon Nanofiber Aerogel Converted from Bacterial Cellulose for Kilohertz AC-Supercapacitors. *MRS Adv.* **2018**, *3*, 855–860. [CrossRef]

62. Beeby, S.P.; Tudor, M.J.; White, N. Energy harvesting vibration sources for microsystems applications. *Meas. Sci. Technol.* **2006**, *17*, R175. [CrossRef]
63. Fan, F.-R.; Tian, Z.-Q.; Wang, Z.L. Flexible triboelectric generator. *Nano Energy* **2012**, *1*, 328–334. [CrossRef]
64. Harne, R.; Wang, K. A review of the recent research on vibration energy harvesting via bistable systems. *Smart Mater. Struct.* **2013**, *22*, 023001. [CrossRef]

© 2019 by the authors. Licensee MDPI, Basel, Switzerland. This article is an open access article distributed under the terms and conditions of the Creative Commons Attribution (CC BY) license (http://creativecommons.org/licenses/by/4.0/).

MDPI
St. Alban-Anlage 66
4052 Basel
Switzerland
Tel. +41 61 683 77 34
Fax +41 61 302 89 18
www.mdpi.com

Materials Editorial Office
E-mail: materials@mdpi.com
www.mdpi.com/journal/materials

www.ingramcontent.com/pod-product-compliance
Lightning Source LLC
LaVergne TN
LVHW071954080526
838202LV00064B/6749